古典细节的艺术

古典细节的艺术

[英] 菲利普·詹姆斯·多德 编
齐梦涵 译

images Publishing
广西师范大学出版社
· 桂林 ·

图书在版编目(CIP)数据

古典细节的艺术/(英)菲利普·詹姆斯·多德编;齐梦涵译.
—桂林:广西师范大学出版社,2017.4
ISBN 978-7-5495-9536-5

Ⅰ.①古… Ⅱ.①菲… ②齐… Ⅲ.①古典建筑-建筑设计-细部设计-文集 Ⅳ.①TU-091

中国版本图书馆CIP数据核字(2017)第031119号

出 品 人:刘广汉
责任编辑:肖 莉 齐梦涵
版式设计:吴 迪

广西师范大学出版社出版发行
(广西桂林市中华路22号 邮政编码:541001
网址:http://www.bbtpress.com)
出版人:张艺兵
全国新华书店经销
销售热线:021-31260822-882/883
恒美印务(广州)有限公司印刷
(广州市南沙区环市大道南路334号 邮政编码:511458)
开本:787mm×1 092mm 1/12
印张:$21\frac{1}{3}$ 字数:50千字
2017年4月第1版 2017年4月第1次印刷
定价:256.00元

"When we build, let us think that we build forever."
John Ruskin

目录

项目集

版权信息

Jim Steinmayer 1992

序

大卫·伊斯顿（David Easton）

在这篇序言中，我想进行一场回顾过去展望未来的古典细节艺术之旅，每位读者都将不可避免地意识到“旧秩序已经改变，新秩序正在形成”这一理念。本书以广阔的视野审视了我们居住的房屋，及其如何与“古典细节”联系起来。

在美国长大的我们对“甜蜜的家”的印象是那些可以使我们联想起往日生活的房屋和当时的生活方式，也就是19世纪中期的那种环境。布杂艺术运动和对古典柱式的应用深刻地影响着美国的建筑学理论，以及我们所建造的房屋。我们从英国、法国和意大利吸收了这种经验，可以说，我们的英国同行对我们“帮助”了很多，他们对古典主义原则的应用已经十分娴熟，而存在于英国和欧洲大陆的辉煌建筑便是最好的证明。

总结过去的十多年，我所提及的观点既追溯过去，又面向我们在21世纪所倡导的未来生活。

我所受的教育首先扎根于普拉特学院的包豪斯运动，但我也有幸在毕业那年夏季获得到枫丹白露学习的奖学金和周游欧洲的机会，切身观赏和感受欧洲的古典建筑。这一去，我完全被古典建筑征服了。对于我们这一代人来说，旅行使一切成为可能。古希腊与古罗马首先使用古典柱式，此后，维特鲁威（Vitruvius）和帕拉迪奥（Palladio）传承了这种建筑风格，这种美妙的继承使古典柱式再次复兴起来，比例、大小及将其加以应用的系统都在从文艺复兴时期到当代的建筑中扮演着重要的角色。

无论在意大利、英国，还是在法国，过去的建筑师将这些原则加以应用，锻炼了我们的头脑和眼睛，使我们能够分辨出什么是古典建筑。在这当中，细节是最要紧的，因为正是通过这些细节，我们在各自领域所受的古典主义教育和进行的古典主义实践才得以彰显，古典主义艺术才得以传承。在我的职业生涯中，我很幸运的获得了大量的旅行机会，也有幸拥有愿意建造带有古典气质的住宅的客户。在这个过程当中，我能接触到许

ABOVE
A rosette, acanthus leaves, and egg-and-dart moldings are some of the decorative Georgian motifs that are included in this wall panel.
David Jones Architect

OPPOSITE
A design rendering for a formal living room
Design by David Easton, and rendering by James Steinmeyer

OPPOSITE
Craftsmen under-cutting the egg-and-dart molding on an ornate plaster crown, that includes intricate acanthus brackets, and rosettes set within the coffers
Foster Reeve & Associates

BELOW
A fluted Doric column with a modillion block entablature
John B. Murray Architect

多受过古典主义训练的建筑师，无论你们相信与否，他们都是用手绘制这些细节设计的，这其中包括石膏天花板、楼梯扶手，以及复杂的挑檐。我非常幸运地拥有乔·马里诺（Joe Marino）这位同事，他向我介绍了很多有关古典细节和绘制技巧的事情。在理解怎样在纸上将这些细节画出来的过程中，眼睛和手都有特殊的作用，不过我有时会想，我们是不是在习惯于使用计算机辅助设计之后，逐渐失去了这项技能。

菲利普在这本精美的图书中介绍的建筑师和手工艺者都对古典细节充满热爱，他们自豪地把这种热爱体现在自己的作品之中。任何热爱建筑的人都会在这些书页中看出使书中所有建筑与众不同的娴熟技艺。书中所有的建筑师都要克服各种各样的问题，以使我们在 21 世纪对项目的要求得到满足的同时，还能让古典细节在项目的结构中发挥应有的作用，他们作品中的美也正是在这个过程中得以体现。

我们现在正以飞快的速度迈进人口激增的 21 世纪，人口统计数据的变化使越来越多的建筑当中不再能看到古典主义的细节。这不是我希望看到的，但是话说回来，我也不能确定未来究竟会如何发展。至少在当下，我们仍然可以惊叹于古典细节的美，并庆幸它们依然保存完好。

Spray

CALIFORNIA GEORGIAN (see page 217): The dining room is decorated with Gracie hand-painted wallpaper that complements the highly detailed door casings and ceiling crown.
Eric J. Smith Architect

FERNE PARK (see page 199): Over the portico of engaged Composite columns, the pediment on the entrance façade is fitted with a carved coat of arms surrounded by flying ribbons, reminiscent of similar designs by Palladio.
Quinlan & Francis Terry Architects

前言

秩序是古典主义建筑师使用的词汇，塔斯干式、多立克式、爱奥尼克式、科林斯柱式和复合式，就连约翰·索恩爵士 (Sir John Soane) 也曾宣称“艺术无法超越科林斯式。”这五种源自古希腊建筑的古典主义建筑形式共同构成了一个有一定秩序的系统，每一种都有其自身独特的比例、细节和特征。没有它们，古典主义建筑就不会产生，而没有对它们的了解，就无法真正了解古典主义建筑。

人们常说建筑在艺术领域是独一无二的，因为它横跨理论和实践两个领域。然而，要想在建筑领域获得成功，不仅需要拥有理论和实践方面的知识，还需要理解和欣赏其建造工艺。建筑师们不会仅凭自己去修筑他们的建筑，他们也要与工匠团队通力协作，通过沟通交流使其理解自己的理念。房屋的风格，所选材料、组成、面积，特别是其中所包含的各种细节，共同构成我们称之为家的建筑。正如温斯顿·丘吉尔 (Winston Churchill) 恰如其分地指出的“我们塑造建筑，而后建筑影响我们。”从这个角度来说，工匠和工艺师对房屋的设计与落成也做出了极大的贡献。

《古典细节的艺术》审视当代古典主义建筑中的优秀案例，聚焦于材料的使用、错综复杂的细节与精湛的技艺，对这些贡献给予了肯定。不同于以往的样式图册，本书旨在为那些对古典主义设计感兴趣的读者提供相关资源。它分为两大部分——论文集与项目集——获得了许多有影响力的学者、工匠与建筑师们的支持，使我们能够读到、看到，及理解他们每一个人在当代古典主义建筑设计中所发挥的举足轻重的作用。

英美建筑师常常从乔治王时代 (1720 年—1840 年) 获取灵感和养分。直到现在，它依然是优雅、精致与高品位的典范。尽管古典主义建筑不是单一的历史风格，认识乔治亚风格及其多个变体在英美两国家居建筑当中所起的作用仍然具有十分重要的意义。这其中比较重要的几种风格为: 以罗伯特·亚当 (Robert Adam) 和詹姆斯·吉布斯 (James Gibbs) 为代表的具有新古典主义设计风格的乔治中期巴拉迪欧风格、以约翰·纳西 (John Nash) 和约翰·索恩爵士为代表的乔治末期摄政风格和乔治亚殖民风格，后来还出现带有帕拉迪奥元素，并借鉴了罗伯特·亚当的设计和古代庙宇的美国联邦风格。大西洋两岸的工艺技术在这期间达到巅峰，这在很大程度上可归因于本书的同类书籍，书中收

ABOVE
Carhart Mansion is the first classical building to be built in Manhattan since the 1960s.
John Simpson & Partners

OPPOSITE
Fluted pilasters with delicately detailed Corinthian capitals
Peter Pennoyer Architects

BELOW
A GEORGIAN MANOR HOUSE (see page 181): Precisely carved from limestone, the door surround and broken pediment, are in contrast to the rustic dry-laid fieldstone walls, which aids in highlighting the importance of the entrance.
John Milner Architects

OPPOSITE ABOVE
Apprentices at SYMM's workshop are trained in the traditional skills of specialist joinery and cabinetry, finish carpentry, painting, decorating, and polishing.

OPPOSITE BELOW
This bespoke limestone fireplace showcases the collaboration between the client, who provided a photo taken on their travels, the architects who then modified the design to fit within the aesthetic of the home, and Traditional Cut Stone were responsible for the carving and fabrication.
Wadia Associates

录的一流建筑师设计的引人入胜的建筑图纸必然为当时的设计师和建造者提供了丰富的素材与启迪。从整体设计到有关比例、细节和装饰物的理论，这些配有插图的作品集涵盖了与此种建筑风格有关的一切知识，成为当时从伦敦到费城每一名建筑师、建造者、砖匠、木匠、泥瓦匠和石膏师的必备读物，直到现在依然被本书中出现的许多建筑师和工艺师援引。在本书的第一篇文章《创造自己的田园生活》（第 31 页）中，新闻记者及历史学家杰里米·穆森（Jeremy Musson）进一步探讨了我们对一切乔治亚风格事物的喜爱，将古典主义住宅描述为塑造幸福美满的家庭生活的完美媒介，而这一主题也贯穿于本书的其他文章之中。

古典这个词的概念被用在人类努力的各种领域中，表示卓越的涵义，然而其最狭义的解释仅表示伯里克利时期。在《古典建筑的语言》（第 35 页）这篇文章中，詹姆斯·斯蒂文斯·柯尔（James Stevens）教授探讨了应用于语言和建筑上的古典理想。著名历史学家及作家大卫·沃特金（David Watkin）在《音乐与建筑》（第 41 页）一文中着眼于这两个不同的学科，并重点比较它们在基本原理、和谐及有序化系统等方面相似的使用方式，这些用法扩散于整个古典主义之中。这些明确的英式解读可以追溯到建筑师伊尼戈·琼斯（Inigo Jones）（1573—1652），正是他建立起帕拉迪奥在《建筑四书》（Four Books of Architecture）中描写的古典主义的规则。罗伯特·查塔姆（Robert Chitham）在他的文章《关于数学和对称》（第 49 页）中讨论了帕拉迪奥母题及对其细节的应用，以及维屈鲁浮斯（Vitrivius）、莱昂·巴蒂斯塔·阿尔伯蒂（Leon Battista Alberti）和塞巴斯蒂亚诺·塞利奥（Sebastiano Serlio）等人提出的比例理论。著名历史学家亨利·霍普·里德（Henry Hope Reed）的文章《古典装饰》（第 55 页）为第一部分文章收尾，再次提醒我们装饰和细节给了房屋风格与身份认同。

在作家与学者的文章后面，是一系列来自英美著名建筑师和设计师的文章。以引用温斯顿·丘吉尔（Winston Churchill）的名言“一个逃避、忘记过去的民族根本没有未来”开始，昆兰·特里（Quinlan Terry）在《逆流而上》（第 59 页）中对当下的建筑状态给出了自己的专业见解。

虽然本书中的大多数材料与历史有关，但是这些内容亦关乎未来。具体来说，就是我们的传统在现在和未来所担当的角色，罗伯特·亚当与休·彼得（Hugh Petter）分别在《传统与创新》（第 65 页）与《工艺的未来》（第 91 页）中对这个主题进行了探讨。朱利安·比克内尔（Julian Bicknell）与约翰·辛普森（John Simpson）在《古典建筑细部的工艺》（第 69 页）与《屹立不倒的建筑》（第 75 页）中延续了对传统的充满热情的讨论，尤其是其与工艺及建造技术的关联。建筑师与作家彼得·彭诺耶（Peter Pennoyer）撰写的《古典建筑中的阿拉伯式花纹》（第 79 页）探讨了伊斯兰教几何图案与古典细节的混合产物，这种用法可以追溯到那些在镀金时代进入法国国家高等美术学院学习的人。学术研究的主题又出现在《设计开发与分析》（第 83 页）和《完整绘制》（第 87 页）当中，在这两篇文章中，理查德·卡梅伦（Richard Cameron）和约翰·默里（John Murray）热切地写下了绘画教育与建筑实践的重要性。

最后，艾登·莫蒂默（Aidan Mortimer）、阿兰·奥利弗（Alain Olivier）、保罗·切斯尼（Paul Chesney）、凯文·克罗斯（Kevin Cross）、理查德·卡宾诺（Richard Carbino）、福斯特·里夫（Foster Reeve）和吉姆·马修斯（Jim Matthews），七位世界一流建筑工艺师讲述了他们所处特殊工艺领域里的情况——包括细木工、金属加工、壁炉、石雕、装饰画与镀金画框、装饰石膏，以及手工砖——以及他们在当今住宅的成功设计中的持续的重要性。

本书的第二部分项目集呈现了 25 个优秀的古典风格住宅，这些当代建筑作品运用了前文提及的各种理论，展现出英美两国著名建筑师的一流设计。这些建筑师包括明显受到伊尼戈·琼斯（ Inigo Jones ）和帕拉迪奥式建筑影响的朱利安·比克内尔（Julian Bicknell）、费尔法克斯和萨蒙斯事务所（Fairfax & Sammons）、彼得·彭诺耶（Peter Pennoyer）、吉尔·谢弗（Gil Schaffer）和昆兰·特里（Quinlan Terry）；作品根植于乔治亚和新乔治亚设计原则的艾伦·格林伯格（Allan Greenberg）、约翰·米尔纳（John Milner）、艾里克·史密斯（Eric Smith）和蒂尼亚·瓦迪亚（Dinyar Wadia）；以及集中体现了索恩和纳西的摄政风格的约翰·辛普森（John Simpson）和弗朗西斯·约翰逊（Francis Johnson）的工作室。弗兰克和洛森事务所（Franck & Lohsen）的作品体

ABOVE
A stylized plaster modillion with a fluted and beaded overdoor detail
John B. Murray Architect

OPPOSITE
A CLASSICAL POOL PAVILION
(see page 145): Ionic column capitals and a Greek key frieze are carved from limestone.
Curtis & Windham Architects

现了范布勒(Vanbrugh)和豪克斯穆尔(Hawksmoor)的英国巴洛克式风格;罗伯特·亚当提供了一种更有时代感的古典风格;蒂莫西·布莱恩特(Timothy Byrant)、约翰·默里和安德鲁·史克曼(Andrew Skurman)都在一间城市公寓中展现出了他们的卓越技术。理查德·卡梅伦、柯蒂斯与温德姆事务所(Curtis & Windham)、弗格森与沙玛米安事务所(Ferguson & Shamamian)、罗伯特·富兰克林(Robert Franklin)、大卫·琼斯(David Jones)、休·彼得、杰夫·史密斯(Jeff Smith)和肯·塔特(Ken Tate)则使这个杰出建筑师与设计师的群体更加丰富。

虽然书中的大多数资料都引自过去,但是这本书并不是一本给设计师提供精确规则,让他们严格遵守的范本或建筑手册。正相反,它是对目前的古典理论、工艺和实践情况的调查成果。本文不可能是一个面面俱到的参考资料,正因如此,我们在参考这一部分(第243页)提供了进一步的阅读参考书目。

由维特鲁威最先提出的秩序的概念贯穿于这些书页所包含的全部作品中。亨利·沃顿爵士(Sir Henry Wotton)在1624年出版的《建筑学要素》(The Elements of Architecture)——这是维特鲁威的《建筑十书》(de Architectura)的第一个英译本——中有这样一句话,"良好的建筑有三个条件:方便、坚固、愉悦。"鉴于现代建造技术与土地规则的发展,所有新建房屋都已经能够达到结构良好且满足基本实用功能的要求。但是并不是全部新建房屋都能产生感官上的愉悦。而为了达到这个目的,建筑学,即建造的艺术,需要立足于传统,这只能靠研究其他目的相同的设计和使用古典建筑语言来实现。《古典细节的艺术》一书中介绍的作品展示了古典主义永恒的美,其精美的细节和高超的技艺在创造性艺术中扮演着重要的角色。

OPPOSITE
FARMLANDS (see page 149): The Serlian window in the dining room includes Ionic columns, and a keystone that engages with the room's crown. *Fairfax & Sammons Architects*

ABOVE
PIETRA MAR (see page 221): A Roman dolphin-themed capital, carved from Coquina stone *Smith Architectural Group*

OPPOSITE
Hand-forged from mild steel, the design for this balustrade in a Palm Beach residence by Smith Architectural Group was inspired by the Beauvais Cathedral communion rail. *Gold Coast Metal Works*

ABOVE
Inlaid with a Greek key pattern, the frieze of this Georgian chimneypiece is framed by consoles, delicately carved out of statuary marble. *Chesney's*

ABOVE
HENBURY HALL (see page 133): This plaster crown includes a floral scroll within the frieze and stylized seashells on the ovolo.
Julian Bicknell & Associates

OPPOSITE
A fluted Roman Doric column with an entablature that is embellished in-between the guttae, on the soffit of the corona.
Allan Greenberg Architect

"A building without ornament is like the heavens without stars."
George Santayana

论文集
THE ESSAYS

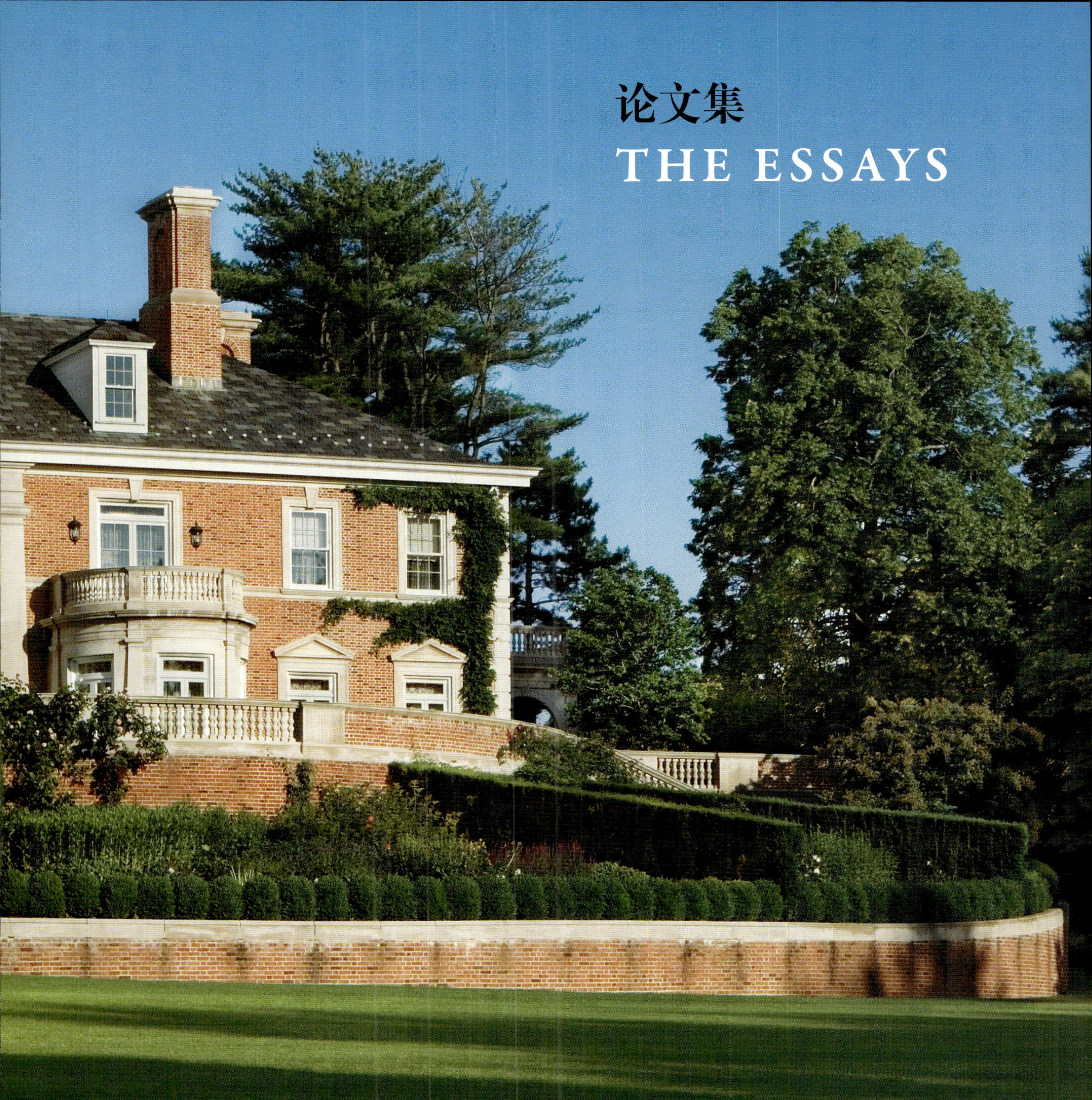

Although at first glance Mickley Park in North Yorkshire appears to be an authentic Georgian home with period details, it also incorporates modern construction techniques and innovative green technologies.
Francis Johnson & Partners

Creating the Private Arcadia

Jeremy Musson

创造自己的田园生活

IF I HAVE OBSERVED one thing in all my years of writing about Classical country houses, it is that they were even in the 17th and early 18th centuries considered to be the perfect medium for a happy and fulfilled life by their creators—and both patrons and designers were creators then. Indeed today, this is also often the case, for this ideal is, I think, reflected in the many contemporary Classical houses I have visited in England and in the United States of America. In these recent houses the emphasis is on creating a harmony and a sense of order and beauty in which detail is used with restraint and discipline. These houses are more usually drawn from the model of the later-18th-century Georgian house and are, in spirit, dedicated to the private family life of the patron as a contrast to the chaos and discordance of life outside the walls of the house and the boundary of the park.

It could be argued that the Classical interior is today, in essence, a private arcadia—an interior paradise—created for the patrons and their children and generations yet unborn, as the secure setting and the shelter for a happy life. This would be a direct inheritance from the spirit of 18th-century interiors. The original arcadia as expressed by the poets of the Ancient world was an idealised and remote pastoral community to belong to which led to simple contentment; the arcadia of the contemporary Classical family home is in the same way an idealised place within a landscape which binds all its inhabitants in bonds of love.

The inspiration of the classical world is recognised in the notes made by Nathaniel Scarsdale, the patron of Adam at Kedleston Hall in Derbyshire, when he wrote in his notebook as he embarked on his great project to create a suitable seat for his family in the Classical style, "Grant Me, ye gods, a pleasant seat, in Attic elegance made neat" after which he goes on to celebrate the joys of family and friendship that would be enjoyed within his architectural achievement. Order and joy "in Attic elegance made neat."

In the early 18th century, the "private arcadia" was usually reflected vividly in the iconography drawn from Classical mythology, especially as translated by the British poets and poured into the Palladian interiors by William Kent and Lord Burlington in painted decoration and plasterwork, in marble, and in stone, as well as on tapestries, paintings, and details on carved and gilded furniture as well as the sculptures and—those outdoor living rooms of the Georgian gentleman—the Classical temples in the park.

The historic iconography of classical mythology was re-introduced into the consciousness of the educated man during the renaissance and put into vivid English by poets led by Alexander Pope. Within these stories lay a wealth of dramatic accounts of the encounters between the mortal and the deity, which seem to elevate the nature and status of the architectural space they decorate. The activities of the loving deities as described in Ovid's *Metamorphoses* are encountered over and over again in the interior decoration of the country house, and the theme depicted is so often that of love or the triumph of love, or the triumph of love and the return of spring.

Of course, such scenes were chosen by patrons and decorative artists and sculptors for decorative reasons, but they were not chosen for decorative reasons alone.

Apollo, the deity of the sun and of music and poetry, is often seen presiding over the principal rooms of the house—not least as the god of light and warmth, but also of music and poetry. Bacchus, the god of wine and feasting, Ceres, the goddess of the harvest, and Venus the goddess of love all appear in their appropriate places. They can appear in combination for as the Latin poet, Terence, had it, "Sine Ceres et Libero friget Venus," without wine and feasting love grows cold.

It may be observed that there only hints towards these images in the contemporary Classical interior and, more often, they are perhaps merely implicit in the architectural details. But they are there. Apollo is there in all the lightness of the interior, Ceres and Bacchus in the elegant proportion of the dining room—perhaps Venus is found within the bedroom? More explicitly the Classical urn is drawn from antiquity and echoes the Ancient Classical tradition of honouring ancestors—as urns held the ashes of ancestors—in the ancient world a man who had no honour for his ancestors had no honour for his family or himself.

Any pediment encountered today, within or without, is also nod to the pediment of a temple dedicated to the classical deities, the gods who in the Classical imagination gave love to the mortal world. The Classical Orders with their individual characters—masculine Doric, graceful young woman in the Ionic, and the full grown matron in all her beauty represented by the Corinthian, and their associated language of details—are also eloquent in the spirit of our shared Classical past. Light, symmetry, order, and discipline can all be found in the well-designed contemporary minimalist architecture too—it is in the detail and proportion that contemporary Classicism tells its own story.

The well-judged employment of Classical detail in a new home has an additional significance that cannot be underestimated. It is an expression of an informed personal choice and an aesthetic statement of faith in civilized values and an evocation of the delight in the human senses—light and touch, the sensation of the warmth of the sun falling on stone. This is true of all the houses featured in this book. The hand-crafted capital and cornice are symbols of love and optimism in a turbulent world. They are above all, things of simple beauty which speak of one well-tried and well-loved version of human dignity expressed in architecture.

ABOVE
A GEORGIAN COUNTRY HOUSE (see page 165): The bespoke wood fireplace in the dining room references the Ionic columns found in the entrance hall, and includes chinoiserie-inspired motifs that complement the hand-painted 18th-century Chinese wallpaper.
Allan Greenberg Architect

OPPOSITE
A GREEK REVIVAL COUNTRY HOUSE (see page 203): Typical of many classical designs, the staircase passes in front of a window to maintain symmetry at the front façade of the house.
G. P. Schafer Architect

WAYSIDE MANOR (see page 157): The sun-filled gallery links all the principal rooms on the first floor of the house, and is detailed with bronze French doors from Gold Coast Metal Works, and engaged Ionic columns carved from limestone.
Franck & Lohsen Architect

The Language of Classical Architecture

James Stevens Curl

古典建筑的语言

FIRST OF ALL, a few definitions might be useful. The term "classical" implies something of the first rank, highest class, order, authority, or importance, a standard, or a model.[1] It is therefore exemplary, but in the present context it specifically suggests that which conforms in style or composition or both to the rules or models found in the Architecture of Graeco-Roman Antiquity.

The Classical ideal is characterised by clarity, completeness, symmetry, deceptive simplicity, repose, and harmonious proportions, and is associated with civilised life, perfection, taste, and serenity. The Classical Language of Architecture[2] is not a free-for-all in which elements are arbitrarily thrown together; it is a highly sophisticated system that was in use throughout the Hellenistic World and the Roman Empire, and its traces survived throughout the centuries, especially in Italy, Greece, parts of France, and the Middle East. Classical Architecture is permeated with elements that can be related directly to the architectural *vocabulary* and *language* of Graeco-Roman Antiquity (especially those found in Greek temples and the ceremonial, monumental, and public Architecture of Ancient Rome), or are derived as variants of such vocabulary and language: those elements include the Orders themselves (consisting of bases, column- or pilaster-shafts [fluted or plain], capitals, and entablatures, in other words variations on columnar and trabeated forms of construction), the surroundings and heads of window-, door-, or other openings (such as niches), pedimented gables, plinths, crowning cornices, string-courses, a very large range of mouldings, with or without enrichment, and much else.

And what of "Language?" A spoken language is a system of vocal symbols by means of which members of a social group participating in a culture interact and communicate with each other. It is the whole body of words and of methods of combinations of words that comprise a "tongue," and is capable of expressing the thoughts, feelings, wants, and so on of human beings. Language in a generalised sense consists of words and the methods of combining words for the expression of thought, and it can also be the phraseology or terms of a science, art, specialism, etc., including Classical Architecture: it has *meaning*, and so "classical," signifying the conforming to rules or models of Greek and Latin Antiquity, when applied to "Language" and "Architecture", suggests something coherent, not meaningless, with all sorts of allusions, capable of enormous ranges of expression (like the spoken language of a culture). And Classical Architecture has a rich language, full of nuances of meaning, that was understood by the cultures which gave birth to it and by those who rediscovered it.

True Architecture can engage the mind as well as shelter the body, but when the base-shaft-capital-entablature that comprises an Order of Architecture is considered, obviously it is not merely a decorative element applied to a building for aesthetic reasons: it is actually a satisfying expression of structure and stability, important psychologically as well as physically to those who behold it. It may appear extremely robust and powerful, or it may be elegant and highly decorative and festive, yet in all its guises it actually *looks* stable, inducing confidence in the beholder, and expressing, even emphasising, the columnar-trabeated essence of the structure.

Several scholars[3] have written eloquently of the meaning and origins of the Orders: in Ancient Greece, for example, where three Orders evolved (Greek Doric, Greek Ionic, and Greek Corinthian[4]), it is clear from the writings of Euripides (*c.*484–407/6 BC) and others that columns of an Order had anthropomorphic resonances. In *Iphigenia in Tauris*, for example, a column is associated with Iphigenia's brother, Orestes, and columns are equated with the sons of a House. This connection is not difficult to understand, because the regularly-spaced rows of columns in a colonnade suggested the lines of soldiers in a disciplined military force, associated with Manliness, Strength, Order, and Stability. In short, we cannot ignore the "talismanic property"[5] of the column. Vitruvius (*fl.* later 1st century BC) suggested which Orders of Architecture might be appropriate for temples dedicated to certain deities,[6] and indeed the Ionic Order's femininity is emphasised by the six *caryatides* supporting the entablature of the south porch of the Erechtheion in Athens. Maleness, on the other hand, was suggested by *atlantes* (as on the Greek temple of Zeus Olympius, Argraces [or Agrigentum]) and *telamones*.[7] But in addition to the three Greek Orders (and their Hellenistic variants, of which there were many), there were the five Roman Orders (Tuscan, Roman Doric, Roman Ionic, Roman Corinthian, and Composite, three distinctly different from the Greek versions and two unknown in Grecian Architecture), and the Roman Orders were often applied (engaged) with walls, mixed with arcuated elements, and transformed in the process. Of course, the canonic Five Orders of Roman Architecture were revived from the Renaissance[8] period onwards, and the Greek Orders re-appeared from the 18th century, so there are eight Orders in all, with many variations, and these, with a vast range of details, mouldings, etc., constitute a huge resource, not only an alphabet, but a vocabulary and a marvellously flexible and universal language.

In several countries today, Classicism in Architecture is alive and well, although it gets a bashing in the Public Prints from time to time, usually by those for whom it never had relevance on any level. Yet visitors to old cities in Italy are surrounded by Classical Architecture (which is almost the vernacular style there): Rome, for example, is a Classical city *par excellence*. Classicism is dominant in Paris and other French cities, and, in its many transformations, can enchant those who trouble to explore (for example) the exquisite 18th-century Rococo churches in Southern Germany. Many public buildings of the 19th and 20th centuries, familiar to multitudes on both sides of the Atlantic, are fine Classical compositions, and it is reasonable to claim that the Classical Language of Architecture contributes not a little to the pleasures of travel, even if it is only manifest as a shadow or paraphrase lurking behind the most stripped and bare essays in unornamented design.

Looking back over more than two millennia, it is clear that the Classical Language of Architecture has enabled the creation of great buildings and even whole, harmonious blocks in cities because it is infinitely adaptable, enabling rich invention and infinite

Architectural Composition I: This analytique assembles the details of the Corinthian order from Zappeion Hall, Athens.
Oil on canvas by Carl Laubin

A PARK AVENUE APARTMENT (see page 137):
The dining room is the apartment's exclamation mark, where highly embellished moldings complement the specially commissioned Chinese murals, and collection of antique Irish furniture.
Timothy Bryant Architect

numbers of variations on its basic themes to blossom. Modernism, however, has discarded a great language, a mighty and expansive vocabulary, and a whole system of grammar and syntax, in favour of a series of monosyllabic grunts, with the predicatable making of a ghastly dystopia of uninhabitable cities, incessant noise, and violent pornographic "entertainment": so, without an alphabet (only a few images of approved exemplars), a vocabulary, or a language, Modernism only represents itself and Barbarism. Now we have the flimsy codification known as Deconstructivism, with its distorted, misshapen, and menacing buildings, hailed by some as a new paradigm, but in fact devoid of language other than threatening assaults on perceptions, serenity, and repose, from which all anthropomorphic resonances have been expunged, hence the unease it induces in the beholder.

The *tabula rasa* demanded by Modernists jettisoned a great language, with all its possibilities, in favour of fancy wordplay, empty jargon, meaningless sound bites, and the sloganising cacophony accompanying a bogus linguistic dance. It has been and is promoted by self-serving, self-regarding totalitarians interested only in power and money, and is ingested by a terrified, bemused, bovine public, too cowed and ill-educated to be able to protest or resist. One of the big problems is that a visually illiterate and desensitised public can only look with its ears now, and confuses obfuscation for profundity; cults invent their own languages and liturgies, the more obscure and dark, the better they serve their protagonists.[9]

Once, in the cubicle of a university lavatory, I saw some graffiti which summed it all up. The text read: "I have just had a Modern Movement." At least somebody in that "dark Satanic"[10] mill was perceptive. It is significant that a mind as receptive and perceptive as that of Johann Wolfgang von Goethe (1749–1832), according to Johann Peter Eckermann (1792–1854), could describe Classicism as Health, but Romanticism as Disease.[11] It would be interesting if he were able to comment on the situation today.

1. Murray, James A. H., Bradley, Henry, Craigie, W.A., & Onions, C.T. (eds.) (1933): *The Oxford English Dictionary* (*OED*) (Oxford: The Clarendon Press), p. 467.
2. Summerson, Sir John Newenham (1980): *The Classical Language of Architecture* (London: Thames & Hudson).
3. Hersey, George (1989): *The Lost Meaning of Classical Architecture: Speculations on Ornament from Vitruvius to Venturi* (Cambridge, MA: MIT Press), or Onians, John (1988): *Bearers of Meaning: The Classical Orders in Antiquity, the Middle Ages, and the Renaissance* (Cambridge: Cambridge UP), for example.
4. Though many variants on these three occurred in the Greek World, notably in the Hellenistic colonies in Asia Minor, as can be studied with benefit in, for example, the marvelllous Pergamon Museum in Berlin.
5. Onians (1988), p.8.
6. Vitruvius Pollio, Marcus (1999): *Ten Books on Architecture.* Translated by Ingrid D. Rowland, with commentary and illustrations by Thomas Noble Howe, and additional commentary by Ingrid D. Rowland & Michael J. Dewar (Cambridge: Cambridge UP), passim.
7. See Curl, James Stevens (2006): *A Dictionary of Architecture and Landscape Architecture* (Oxford: Oxford UP), pp. 50, 155, 359, 765, 769 for definitions
8. See Alberti, Leon Battista (1988): *On the Art of Building in Ten Books* . Translation of *De Re Ædificatoria* (1486) by Joseph Rykwert, Neil Leach, & Robert Tavernor (Cambridge, MA, & London: MIT Press).
9. Salingaros, Nikos A., with Christopher Alexander, Brian Hanson, Michael Mehaffy & Terry M. Mikiten, and a Foreword by James Stevens Curl (2004): *Anti-Architecture and Deconstruction* (Solingen: Umbau-Verlag), pp. 15–16.
10. Blake, William (1804–10): *Milton: a Poem,* etc. (London: William Blake) Preface, *And did those feet...*, line 8.
11. Eckermann, Johann Peter (1836–48): *Gespräche mit Goethe in den letzten Jahren seines Lebens, 1823–1832* (Leipzig: F. A. Brockhaus), 2 April 1829 *(Das Klassische nenne ich das Gesunde, und das Romantische das Kranke).*

HOLLYWOOD REGENCY (see page 153): The great room features walnut bookcases and paneling, as well as a coved ceiling.
Ferguson & Shamamian Architects

Music and Architecture

David Watkin

音乐与建筑

MUSIC AND ARCHITECTURE[1] are linked at the dawn of our civilisation by the celebrated musician, Amphion, who features in the writings of Homer and Euripides.[2] Son of Jupiter and Antiope, Amphion built the walls of the city of Thebes by drawing the stones into place with the magical sounds of the golden lyre, given him by Hermes.[3] A beautiful relief of him from the 2nd century AD was incorporated into the Meridiana Gallery at the Palazzo Spada in Rome in the mid-16th century.[4] In the chapter "The Education of the Architect" in *De architectura*, Vitruvius insisted that the architect should "understand music so that he is conversant with the system of harmonic relationships and mathematical theory,"[5] and devoted a subsequent chapter to "Harmonics."[6] The analogy with music is not only mathematical as it is in Vitruvius who was influenced by the Pythagorean theory of harmony as "heard arithmetic." The analogy is also, as we shall see, similar to tonality in music, providing a frame within which part answers part.

In the classical revival of the 18th century, much thought was given to the relation of musical and architectural harmony, notably by the Abbé Laugier, Père Castel, and Le Camus de Mézières. In his influential *Essai sur l'architecture* (1752), Laugier argued that "Round mouldings are to architecture what consonances are to musical harmony, whereas square mouldings correspond to dissonances. The blending of mouldings and sounds has the same aim and must follow the same rules. The round mouldings give all the softness, the square ones the harshness ... Hence the rule not to employ any dissonance which is not prepared and rescued [sauvée] by a consonance."[7] This is a development from Greek musical theory, as summarised by Vitruvius, which divided intervals into those that were concordant and blended together (the octave, fifth, and fourth, as well as those compounded from them); and those (all the others) that were discordant.[8] In his edition of Vitruvius, Berardo Galiani imputed "the degeneracy of modern architecture" to the ignorance of music shown by current architects, in contrast to the ancients who "attended to it in proportioning their buildings."[9]

Laugier was followed by the architect and theorist, Le Camus de Mézières, who wrote a sensationalist exploration of form and its expressive value in which he gave an account of Amphion at Thebes. Here, he claimed that "Music, that divine Art which enchants us, has the most intimate connections with Architecture. They have the same consonances, the same proportions.[10] In considering the harmony between proportional systems and musical intervals, he referred to the "marvellous harpsichord of colours" of the Jesuit priest, Louis-Bertrand Castel, with which in 1754 he "gave a concert of colours and at the same time of sounds."[11] John Soane, who was profoundly influenced by Laugier and Le Camus de Mézières, believed that "Mouldings are as essential and important to the architect as colours to the painter".[12]

The great polymath, Johann Wolfgang von Goethe (1749–1832), said that architecture was frozen music, as recorded in *Conversations with Goethe* by Johann Peter Eckermann (1792–1854). The parallel had been drawn earlier by Goethe's friend, Friedrich von

Schnelling, in his *Philosophie der Kunst* (1802–03), while Goethe himself studied musical theory between 1810 and 1815 in the hope of writing a book on music as a parallel to his *Theory of Colours* (1810). Though the idea of music being frozen is hardly an attractive one, the notion that there is some kind of parallel between music and architecture has much to be said for it. This is partly because architecture is appreciated through its form rather than through any expression of a human narrative, and this largely true of classical music, notably chamber music though not opera.

Durand, Professor of Architecture at the Ecole Polytechnique in Paris from 1790–1830, noted that architectural elements "exist within architecture like words in a discourse or notes in music."[13] Indeed, the grammar of the language of the orders is quite as complex as that of musical notation, so that in classical architecture the relation of every element to every other is based on a module that is the diameter of the column at its foot. We can analyse this proportional relationship in the same way that we can analyse a Bach fugue in which the initial exposition or theme—a variant of the module in classical architecture—is heard successively in all the voices.

The parallel is closest in the relation between music and the grammar of the classical orders in architecture. All classical architecture is based on the orders in a system of trabeation in which columns support a horizontal entablature. These elements are divided into three units: thus, the column has a base, shaft, and capital—the base itself consists of three parts—while the entablature comprises architrave, frieze, and cornice. In temples, the roof which this structure supports has a pediment at each end, its three sides forming a triangle. We see exactly this at a Palladian building such as Houghton Hall, Norfolk, by Colin Campbell and James Gibbs, where in the west garden front of 1722–27 the central three bays are fronted with an Ionic portico of four giant Ionic columns supporting a pediment.

The language of the orders has a grammar that is quite as complex as that of musical notation so that the relation of every element to every other is based on a module that is the diameter of the column at its foot. We can analyse this proportional relationship throughout all parts of the west front at Houghton in the same way that we can analyse a fugal composition by Bach in which the initial exposition or theme is heard successively in all the voices.

Each column on the west portico at Houghton is 30 feet high, nine times the width of its diameter at its foot, a proportion customary in the Ionic order where the columns are more slender than in the Doric order. This module also governs the height of the base of the columns at Houghton which is half the diameter of the column. The intercolumniation at Houghton, that is the width between the columns, is three times the diameter of the column. All this makes the length of the portico equivalent to 13 modules. The entablature which surmounts the columns is one fifth of their height, while the 10-foot-high crowning pediment is three times their height. Though these proportions are always interrelated in classical buildings, they are not fixed, for the great Renaissance theorists like Palladio, Scamozzi, and Vignola, all came up with their own personal variations.

In the most magnificent reception room at Houghton, the Saloon, created by William Kent in 1725–31,

CALIFORNIA GEORGIAN (see page 217): Design renderings for the dining room, living room, library, and stair hall show furniture in the rooms which immediately provides a sense of scale.

CALIFORNIA GEORGIAN (see page 217): Typical of Georgian design, the staircase is located at the end of the principal cross-axis in the house, which in turn is centered on large Palladian window at the landing.
Eric J. Smith Architect

the pedimented Ionic chimneypiece is a brilliant variation on the theme of the west portico, showing the startling effect of the variety allowed by classical discipline. Here, the Ionic columns are coupled, one behind the other; the pediment is broken in the centre to allow for an antique bust; the frieze is interrupted by a marble panel carved with Walpole's star and garter; while a striking polychromy is introduced by the use of white Carrara marble contrasting with black marble flecked with gold which Kent also used for the tops of his side tables in this room. All of this variety against a common classical base, including the echoing rhythms of the columns which flank the chimneypiece, breaking forwards and backwards, can surely be paralleled in the music of Bach, born in 1685, the same year as William Kent. Works like his Double Violin Concerto in D Minor were created while Kent was at work at Houghton.

Just as the basic elements of classical architecture—the column, entablature, and pediment—are divided into three units, so in music, the sonata form that appears in symphonies as well as sonatas, is divided into three: an ABA rhythm of theme, development, and return, including recapitulations and codas. The disciplines of architecture and music also have numerous flexible elements which bring variety to the basic core; for example, the Greek Doric column lacks a base just as a sonata does not invariably have only three movements.

Each of the orders which form the core of classical architecture, Doric, Ionic, and Corinthian, incorporate moldings appropriate to them, which not only serve a decorative purpose but can themselves also be decorated. There are essentially five mouldings, cyma recta, cyma reversa, ovolo, cavetto, and dentil, each of which can be ornamented with decoration such as egg and tongue, bead and reel, and waterleaf. This decoration can be compared to the ornaments which feature so strongly in classical music and were described and explained by C. P. E. Bach in a chapter in Part I of his influential and authoritative treatise, *An Essay on the True Art of Playing Keyboard Instruments*.[14] Referring to the principal embellishments, appoggiaturas, trills, slides, turns, and mordents, which he regarded as "indispensable," he stressed that "All ornaments bear a proportionate relationship to the length of the main note, to the speed and to the expression of the music," thus making a parallel with the architectural language of the orders He went on to explain that the ornaments "join notes and enliven them" as well as "give emphasis and accentuation."[15]

The composer Jean-Philippe Rameau, the founder of tonal harmonic theory in works such as his *Theory of Harmony* (1722), was inspired by the claims of the architect Charles-Etienne Briseux in his *Traité du Beau essential dans les arts* (1752) that the simple numerical ratios and proportions of the classical orders in architecture were based on the harmonic and geometric proportions found in music.[16] Even if that is debatable, the complex moldings and their ornaments certainly have to be mastered by the classical architect, just as the composer of music has to operate within a parallel but flexible grammar.

The unfolding structure of a fugue can remind us of a complex building articulated with a giant order rising through two stories, as pioneered by Michelangelo in the Capitoline palaces, threaded through with a minor order echoing the theme of the major order and producing successive layers of depth. The leading ceramic artist, Edmund de Waal, recently described in terms of music the stone façade of his family's former home, the Hôtel Ephrussi in Paris. He explains that making his own drawings to show "the rise and fall of the depth of the windows and pillars," made him see that "There is something musical in this kind of elevation. You take the classical elements and try to bring them into rhythmic life." He describes the "red brick stable block with servants quarters above" as "a pleasing diminuendo of materials and textures."[17]

Adherence to these respective disciplines of classical architecture and music provides a base on which the artistic imagination of the architect and composer can flourish. An illuminating analogy with this process of creative tension was provided by the Cambridge historian, F. A. Simpson, who observed, "how unimaginably fast and far the mind of man can rove when anchored to a liturgy."[18] C. P. E. Bach, Haydn, and Mozart could spin astonishingly lyrical poetry around such a formal framework, incorporating dissonances which make sense through their relation to consonances, as Laugier recommended. This is the opposite of the dissonances in buildings by architects such as Zaha Hadid or Frank Gehry which are meaningless and offensive, grammarless and arbitrary because they are not "rescued by consonances" and contain no details which provide a reason for or anticipation any other. Yet architects as varied as Soane and Lutyens adhered to the classical discipline of Vitruvius while creating uniquely personal classical buildings, as do the architects of today whose work is illustrated in this book.

---

1 I am grateful to Roger Scruton for his helpful comments on the first draft of this essay.
2 Homer, *The Odyssey*, xi 260–5.
3 Euripides, *Phoenician Women*, 824: "the walls and towers of Thebes rose to the sound of his lyre."
4 It is one of eight antique reliefs depicting mythological subjects in La Meridiana. See Lionello Nelli, *Palazzo Spada* (Rome: Editalia, 1975), pp. 189–204.
5 Vitruvius, *De architectura*, I, i. In the same chapter, his reference to "the harmony of the stars", better known to us as "the music of the spheres", was derived from Pythagorean cosmology where the orbits of the planets were seen as placed according to musical intervals.
6 *Ibid.*, V, iv.
7 Marc-Antoine Laugier, *Essai sur l'architecture* (2nd ed., Paris, 1755), p. 68.
8 See Martin West, *Ancient Greek Music* (Oxford, 1992).
9 *L'Architettura di M. Vitruvius Pollione* (Naples, 1758), a point noted by William Newton in his edition of Vitruvius (1771, vol. 1, p. 4), the first English translation of Vitruvius.
10 Nicolas Le Camus de Mézières, *Le Génie de l'architecture; ou, l'analogie de cet art avec nos sensations* (Paris, 1780), p. 11.
11 *Ibid.*, p. 10. This is, of course, related to the connection between the seven notes of the musical scale and colour intervals in Isaac Newton's *Opticks* (1704).
12 MS note dated 12 September 1809 (Sir John Soane's Museum Archives 1/164/6, fol. [1]).
13 Jean-Nicolas Durand, *Précis des leçons d'architecture*, vol I (Paris, 1802), pp. 29, 30.
14 Part I of Bach's treatise was published in 1752, Part II in 1765; William Mitchell, transl. and ed. (New York: Norton, 1949).
15 See Philip Barford, *The Keyboard Music of C. P. E. Bach: Considered in Relation to his Musical Aesthetic and the Rise of the Sonata Principle* (London: Barrie and Rockliff, 1965).
16 Jean-Philippe Rameau, *Treatise on Harmony*, transl. with intro. and notes by Philip Gossett (New York: Dover, 1971). See Thomas Christensen, *Rameau and Musical Thought in the Enlightenment* (Cambridge, 1993), pp. 232–5.
17 Edmund de Waal, *The Hare with Amber Eyes: A Hidden Inheritance* (London: Vintage Books, 2011), p. 21.
18 Frederick Arthur Simpson, *A Sermon Preached at the Commemoration of Benefactors* (Cambridge, 1932), p. 6.

A computer rendering of a new brick and cut limestone English Manor House in Hampshire
Franck & Lohsen Architects

The doorways of the principal rooms in this Palm Beach residence have been given added importance by departing from the typical door casing used elsewhere in the house, and creating an elaborate surround that is embellished with scrolled brackets.
Ferguson & Shamamian Architects

On Mathematics and Symmetry

Robert Chitham

关于数学和对称

ARCHITECTURAL THEORISTS of the Renaissance were united in holding up architecture as a mirror to nature, but were not inclined to spend too much time in amplifying this proposition, or in investigating the nature of beauty, or how classicism responded to it. This is a matter of context and confidence. Renaissance education took in, indeed was directed by, the study of ancient Greece and Rome. Just as language itself was derived chiefly from these sources, so were the arts. Therefore they were not moved to question the unique validity of classicism, but rather to control, codify, and develop the orders that constituted its essential grammar. Most Renaissance parallels of the orders move on rapidly from questions of principle to concern with the actual proportions of columns, entablatures, intercolumniation, and so forth. One or two authorities, notably Alberti, break this pattern by returning more than once to seek answers to what constitutes beauty. But all come to the conclusion that the chief determinants of beauty are symmetry and harmony of proportions.

The earliest writer whose work on architectural theory survives is of course Vitruvius. In his *Ten Books on Architecture*, written in the 1st century BC, Vitruvius identifies six fundamental principles of architecture, of which the first four are entirely concerned with the aesthetics of design. They are *Order* (taxis), *Arrangement* (diathesis), *Eurhythmy* and *Symmetry*. The fifth and sixth principles—*decor* (seemliness) and *distribution* (economy)—concern the appropriateness of the building to its social and physical context. But it is Vitruvius's principle of order that introduces the notion of the module as a basic unit of measurement from which all the elements of a design can be generated. This firmly anchors architectural design to a system of mathematics, a proposition that Renaissance writers acknowledged as the basis of their search for perfect proportions in the orders.

It is not until the early Renaissance in Italy that the architecture of ancient Rome began to be re-explored and analyzed, chiefly as a basis for contemporary architectural design. In this regard, Vitruvius, as the only surviving Roman architectural text, proved an essential starting point for the new theoretical examination of principles.

The earliest, and in some respects the greatest, of the Renaissance analysts, Leon Battista Alberti (1404–72), based his ten books on architecture on those of Vitruvius. Alberti declares at the outset his intention to discover what constitutes the ideal form of a building, drawing on the words and the works of "the Ancients" and supplementing these with his own observations and experience. He states that "the Art of Building consists in the design and in the structure ... It is the property and business of the design to appoint to the building and all its parts their proper places, determinate number, just proportion and beautiful order, so that the whole form of the structure be proportionable." Unfortunately Alberti (even more than Vitruvius) is more concerned with laying down what constitute ideal proportions than attempting to explain why they are ideal. Alberti returns repeatedly to the question of beauty, stating that "it is generally allowed that the pleasure and delight which we feel on the view of any building arise from nothing else but beauty and ornament." He goes on to argue that in building, beauty is of more consequence than utility, implying that the latter is comparatively easy to attain, before defining beauty as "a harmony of all the parts, fitted together with such proportion and connection that nothing could be added, diminished or altered, but for the worse." Therefore, he says, that part of design "which relates to beauty and ornament... must without doubt be directed by some sure rules of art and proportion." These rules, he believes, "were begot by chance and observation, and nursed by use and experience, and improved and perfected by reason and study."

In his four books, Andrea Palladio (1508–80) does not enter into a discussion of beauty and what are its architectural attributes, but asserts that "since architecture imitates nature (as do all the other arts), it cannot endure anything that alienates and distances it from what nature herself permits." He believes that many architectural elements are proportioned as they are to ensure that they appear natural, and for example compares the diminution of the column with the tapering of a tree trunk. Overall, Palladio's four books are characterized by a reluctance to be drawn into too much theory—he blames Alberti for being too prolix —and for reliance on illustrations accompanied by comparatively brief text.

The *Five Books* of Sebastiano Serlio (1475–1554), published from 1537 regard geometry as "the first degree of all good art" and starts his first book, *Of Geometry*, with a definition of basic terms, and a reprise of simple geometrical figures including the golden section. His second book is somewhat surprisingly devoted to perspective. This perhaps reflects his career as a painter before he took up architecture. He justifies this parade of his expertise in perspective by showing how crucial it is in theatre, and specifically, scenery design. Serlio's third book contains an anthology of Roman buildings rather similar to that which Palladio was to publish 35 years later, but packed with much more detail (though fewer dimensions). In his fourth book, Serlio comes to his own version of the five orders, again embellished with masses of detail and suggestions of how they might be employed in practice. But his illustrations are almost entirely without proportional dimensions, which are supplied only in the text.

All these Italian authorities believed that the orders could be defined by a fixed set of proportional dimensions, which were the key to the beauty of the building—though their ideal proportions differed from one another. Serlio was modest enough to accept that even his "perfection" may in the future be improved on.

It is fair to say that classicism is concerned with the deployment of simple shapes' geometry, the circle, the square and rectangle, and the triangle, and their combination in all sorts of ways, some very straightforward, and some of amazing intricacy. And of course, some classical elements come directly from nature, for example the volute which is subject to precise geometrical rules. This is why there has been a continuing tendency for scholars to find complex patterns, particularly of squares and equilateral triangles in places, ever since Serlio concluded his treatise on Geometry with a triangulated diagram of the proportions of a doorway. Any number of these analyses are post-rationalisations, often based on an

The wood floor is painted in a marquetry-like pattern, and the door casing engages with the room's entablature enabling triglyphs to be used to further accentuate and decorate the passage from one room to another.
John B. Murray Architect

over-complicated visualisation of the actual building. Serlio's diagram consists of triangles generated by a square whose dimensions depend on the height of a small plinth surmounting the pediment.

Although it is true that throughout time, Classicists have been concerned with proportions, this does not mean that they are bound by a rigid system of proportion. It is futile to look for an absolute set of dimensions to govern each order that is ever designed. Contemporary classicists will often say that they regard specific parallels of the orders as unnecessary, and that they feel free to ascribe to an order any set of dimensions they prefer. Rules, it is said, are for the obedience of fools and the guidance of wise men, but freedom to adapt and vary proportions must always be the result of a sound knowledge of principles. Some architects certainly do proportion their plans and elevations from an initial diagram of rectangles and triangles, but the attempt to find these figures in completed works is generally only of interest as an academic exercise; what it demonstrates is that any sophisticated and ordered design will inevitably generate systematic geometrical figures in plan and elevation.

What conclusions can be drawn from this brief review of the works of authorities from Roman times and the Renaissance? The essence of classicism is the freedom of boundless variation within an overriding framework of discipline. This tension is what has produced the infinite flowering of invention which has characterized Classical architecture in the past, and can continue to do so, depending on the daring of the architect.

This essay is a shortened version of a longer work by Robert Chitham entitled "On Proportion."

OPPOSITE:
HENBURY HALL (see page 133): A cantilevered stone staircase climbs the entire height of the house and is illuminated from above by ocular windows and the cupola, all located on the dome.
Julian Bicknell & Associates

BELOW:
The intricate wood carved details of the entrance surround are in contrast to the rustic qualities of the fieldstone walls.
John Milner Architect

The entrance hall of this Palm Beach Italianate style home includes intricate carved stone and a bronze and glass door.
Smith Architectural Group

Classical Ornament

Henry Hope Reed

古典装饰

IT MAY APPEAR UNNECESSARY at this late stage in our history, but we have been forced by an overwhelming opposition to define the classical. What is it, after all, that we are talking about?

We see the classical tradition as a generalized and idealized interpretation of nature, first begun by the Greeks and the Romans and given a new life in the Renaissance—a renaissance, by the way, which lasted in America into the 1930s. In art, style is identified by ornament. A room may be any shape but its ornament is what gives its style and identifies it as to style.

Let me illustrate the point by a common fallacy perpetrated by the art historian. The art historian defines a certain American style of domestic architecture as the Shingle Style. Shingles do not make a style, no material does. Buildings of shingles can be in any number of styles. Underscoring our definition we also say that the Classical style is identified by classical ornament.

There is, or course, a corollary to the definition of the interpretation of natural forms, namely proportion. Certain basic proportions are very much part of the classical, and they are to be found in the study of the Orders. For that reason a key introduction to the classical in architecture is the drawing of the Doric, the Ionic, and the Corinthian.

A classical building cannot be divorced from ornament. Take the case of the Parthenon. The building is a splendid post-and-lintel construction, but it is nothing more than a stand for ornament in the form of abundant sculpture. And if we accept the fact that the glorious temple was once covered with polychromy, we have the role of ornament confirmed more than ever.

Now, in underscoring the role of ornament, or should we say, decoration, let us look at the natural forms which have been generalized and idealized. Of what do they consist? At the lowest level ornament, or decoration, begins with the abstract profiles, the fillet, the bead, the ovolo, the scotia and others so familiar. Next in our hierarchy come the profiles enriched with the egg and dart, the lead and dart, the bead and reel, pearls, and the Vitruvian scroll. Flora stands at the third level. Anthemion, bayleaf, swags of fruit or flowers, and the rosette follow. Above all, there is the acanthus—that blessed link to the Mediterranean. So abundant is the acanthus in the classical panorama that John Barrington Bayley declared it the morphological symbol of Western art, the equivalent of the chrysanthemum in Japanese art. Less abundant, but no less an object of affection are animal forms, which are fourth in our list. Yet another category is that of man-made objects. A favorite is a cluster of musical instruments. At the top of the hierarchy is the human form. In no other artistic tradition does the human body have quite the place of honor that it does in Western art.

Definition and hierarchy may seem a round-about way of accessing ornament in our tradition, but the reward is that we can see the panoply of decoration. Admittedly most of us are limited in executing the lower levels of the hierarchy. In laboring in the classical vineyard the decorator seldom goes beyond floral enrichment. Even today, with the interest in the classical how many architects and decorators are even asked to draw an acanthus leaf? Or how many are free to look at fauna? We appear to be still held back by the fashion for the abstract. Yet there is change. Enrichment is appearing more frequently. The once elusive acanthus is returning; the bayleaf and the rosette will follow. With time, as we move up the hierarchy, the designer will begin to make a place for the human figure, carved and painted.

Of course, you can say that Palladio in his palaces and villas offered the painter nothing more than a flat wall. But it was a flat wall twenty or more feet high. Columns and other devices were conceded by the architect to the artist—who in turn furnished the setting of columns, ornaments, drapery and the rest for the figures. The artist, in a word, executed the frame.

One word about subject matter: today's realism is far removed from that of the 19th century; there seems to be an obsession with the ugly. Ugliness is presumed to be the essence of realism. In the past, if you look at the realism of a Bouguereau you find beauty; in a Sargent, elegance. Today the artist, pre-occupied with himself, seems incapable of executing anything which, to use an old-fashioned word, pleases. Nor should he be afraid of telling a story. We must take as a rule that the artist's duty, as George DeForest Brush observed, "is to paint a noble subject on the wall."

And what is the contribution of the craftsman to all this? He is going to execute the decoration of the rooms, of one or several buildings. At the turn of the previous century he worked solidly in the tradition and, even produced the human figure. Such has been his faith in the traditional that, despite the hardship inflicted by Modern nihilists, he has survived to our time. The craftsmen too, like the rest of us, must seek out the great examples of the past and work with them, for what would be the point of the designer asking for the human form, be it a mask on a keystone or cherubs in a stucco relief, if the craftsman were only capable of attempting the realistic?

We all have to see decoration in terms of the generalized and idealized interpretation of nature. We have to accept the work of the masters of the past imitate when necessary.

Published with permission of the author—adapted from a speech given at "Creating the Classical Interior Today," a conference held in New York on April 24, 1993, and later published in 1997 as part of the "Classical America" newsletter series.

ABOVE
The design for a Ionic capital, inspired by those at the Erechtheion, Athens
Rendering by Francis Terry, Quinlan & Francis Terry Architects

A GEORGIAN COUNTRY ESTATE (see page 231): The bespoke mahogany balusters are designed as Doric columns topped with urns.
Wadia Associates

Juniper Hill, in Buckinghamshire, features a finely detailed giant stone portico.
Quinlan & Francis Terry Architects

Swimming Against the Tide

Quinlan Terry

逆流而上

I SHALL START with a familiar quotation from Edward Gibbon's *Decline and Fall of the Roman Empire*, which many of you know well. He said there were five attributes that characterised Rome at its end.

"First, a mounting love of show and luxury. Second, a widening gap between the very rich and the very poor. Third, an obsession with sex. Fourth, freakishness in the arts masquerading as originality and enthusiasm pretending to be creativity. Fifth, an increased desire to live off the State."

It all sounds pretty familiar!

I shall confine myself to his fourth point—"freakishness in the arts masquerading as originality and enthusiasm pretending to be creativity."

It seems today that we focus on freakishness and originality, and have forgotten the importance of the continuity of traditional architecture in the life of the nation. In the past, every nation, once it had achieved a measure of peace and prosperity, erected monumental buildings in their capital cities which lasted for centuries and were admired by all.

In the centuries BC, Solomon erected the Temple in Jerusalem in the 10th, the Persians built Persepolis in the 6th, the Greeks built the Acropolis in the 4th and the Romans built the Pantheon and the Colosseum in the 1st. And after the return of civilisation, all Europe erected their great Cathedrals. During the Renaissance they erected the buildings we still admire and regard as the essence of both time and place of which Wren's churches, Somerset House, and Versailles are typical examples. Even after the last war, Germany rebuilt the Zwinger in Dresden and Poland rebuilt Warsaw Old Town before they rebuilt their hospitals and schools because traditional Architecture was, to them, an essential part of their national character and identity. We cannot imagine Rome without St Peter's, nor can we imagine London without St Paul's or the Abbey.

All the buildings I have referred to were built in stone, they were symmetrical, beautiful and classical in its widest sense, which would include what we would call Byzantine Romanesque and Gothic, with arches, columns, gables, and pinnacles. That tradition continued with endless variations until about the middle of the last century.

Since then our age—in both East and West—has turned its back on its past and now erects buildings which have no traditional precedents. There has been a change in attitude which de Tocqueville described as a "scission that has taken place in our history." So, for example, the buildings on the London Olympic site are indistinguishable from what one sees in Moscow, Los Angeles, Sydney, Beijing, or Dubai.

The process I am describing is primarily stylistic—traditional versus modernist. But it also affects materials—whereas we used to build in stone, brick, lime mortar we now use steel, cement, plastics, and slabs of glass. It also affects longevity—whereas buildings used to last for centuries now they are frequently demolished in a few decades. A recent American study put the useful life of steel and glass buildings at 25 years. And there is another difference—whereas the old way of building managed without the consumption of oil and electricity, the modern way is wholly dependent on the consumption of fossil fuels not only for construction but also for their daily use, for example in lifts, lighting, and air conditioning.

Now we may have all sorts of reasons why our present predicament has come about, but I simply want to ponder the fact—as we observe the architecture of our present polluted and precarious times which is so proud of the fact that it has no relationship whatsoever with the architecture of any previous generation. People no longer ask. "Is it beautiful? Is it durable? Is it true?" but, "Is it Modern?" Modernism has become the idol of our time that we see so clearly manifested in the Church and in Government as we abandon our Christian heritage. Some describe this changing face of our society and cities as original, evolutionary, innovative, inevitable, and exciting. To others like me it is deeply depressing!

But depression can lead to defiance, which is now overdue. I have always felt, from the moment I started studying architecture over 50 years ago, that if I am ever privileged to be asked to design a new building, I will try to recover that tradition which has been rejected by the Architectural Establishment, all the Schools of Architecture, the Royal Institute of British Architects, the Royal Academy and, when I served on it, the Royal Fine Art Commission.

I have, therefore, had to swim against the tide all my life, constantly ridiculed in the architectural press; never once commissioned by the public sector; and therefore being confined to a few courageous clients, who fortunately, had the means to employ me, so that during my lifetime I have been able to erect over 100 new classical buildings.

I am pleased to say that I am no longer alone in this. Over the last 30 years there have been a steadily growing number of competent architects, developers, and clients, in this country and abroad, who have seen the failures caused by Modernist dogma and are returning to traditional and classical principles. Some of these new buildings are good; some less so, but that is inevitable in a period of transition. Private houses for private clients are now almost exclusively classical. Churches, universities, commercial housing, and offices are a mix of Modernist and Traditional. Major public buildings, sadly, are still the almost exclusive preserve of the Modernists because they are controlled by the mindset of the media and government. It will be interesting to see how things develop—remembering that Churchill said "a nation that forgets its history has no future!"

I suppose, ultimately, the reason why the majority swim with the tide, and a minority swim against it, relates to our world view.

The atheist Richard Dawkins holds the world view that "The universe we observe has precisely the properties we would expect if there is, at bottom, no design, no purpose, no evil and no good, nothing but blind pitiless indifference ... no rhyme nor reason, nor any justice."

If that is our world view, it is bound to affect the way we live and work. We will probably want to make as much money as possible and consume as much as possible, regardless of how this affects other people, because we have nothing better to look forward to.

But that is not my world view.

Here Francis Terry fully renders a drawing of a Corinthian pilaster to show how shadows impact a design.
Quinlan & Francis Terry Architects

Waverton House, in Gloucestershire, is built using local coursed rubble stone and Cotswold stone roof slates.
Quinlan & Francis Terry Architects

A design rendering by Francis Terry of an ornamental panel
Quinlan & Francis Terry Architects

As I observe the universe, I see the opposite of Richard Dawkins. I see beauty, symmetry, order, harmony, amazing mathematical precision, and humility; which convince me that the universe is all the work of the Supreme Architect.

If we think like this it is also bound to affect the way we live. Now, how that works out in practice is open to debate and we all make many mistakes, but to me it would include three things:

In architecture: humility to learn from the works of our forefathers and see how they constructed buildings for thousands of years.

In life: a passion for beauty, symmetry, order, and harmony in all our relationships.

And above all, a desire to know that Supreme Architect better; who is not some man-made god, but the One who became flesh, died and rose again for us, even Jesus Christ our Lord.

Solar House, built in Sussex, takes a modern abstract look at Classicism, while at the same time recalling the neoclassical architecture of Ledoux.
Robert Adam, ADAM Architecture

Tradition and Invention

Robert Adam

传统与创新

CLASSICAL ARCHITECTURE and the detail that defines it are in a constant state of evolution. Even the most elementary historical observation shows that the Greek temple, which lies at the heart of the tradition, has evolved into the basilica, the palace, the town hall and the railway station. The Classical Orders themselves are the product of an evolutionary process: they started with Doric and Ionic, the two tribal Orders; then the Corinthian was added as a luxury form of Ionic; Vitruvius put in the Tuscan to create a fictitious Roman origin; and finally in the Renaissance, trying to make sense of Vitruvius and the variety of survivals from antiquity, the Composite was added. The details that characterise these Orders have also developed. Archaeological evidence shows continuous experiment in Greek antiquity. Roman architects mixed and added to the details they had inherited from the Greeks. In the Renaissance not only were pedestals added to the Tuscan and Doric in order to rationalise the new systemisation of the Orders but two quite new details became permanently associated with classical architecture: rustication and the baluster.

Much contemporary classical architecture is, however, designed on the principle that all composition must be justified by precedent and that this precedent had a *terminus ante quem* in the Renaissance authors who rationalised the Orders, such as Gibbs or Palladio, in the neo-classical period that ended in the eclecticism of the early 19th century, or possibly from the approved historical models in Beaux Arts education. This conservative position can probably be traced back to the isolated position of modern classicists. There is a lack of confidence and a need for reassurance and justification fuelled by a reaction to the cult of novelty of the dominant Modernist architectural establishment, which vociferously opposes any return to classicism.

Ironically, the view amongst the exponents of classical architecture that it is based on fixed canons from the past is shared with its detractors. For the detractors, this makes classical design a phenomenon that belongs specifically to another period and which is therefore alien and inappropriate for the modern world. While there can be nothing wrong in drawing directly from the great examples of the past, it is unreasonable to limit classical design to these precedents. It denies the fundamental aspect of classicism that affirms its continued relevance: that it is *not* history but a tradition. History is a record of something that has happened, is a series of unrepeated sequences and is fixed and factual (even if the facts can be continually reinterpreted). Tradition, on the other hand, is like memory—selective, cross-temporal, additive, didactic, and continually relevant.

Many supporters, practitioners and patrons of classical architecture do indeed see it as a way of recapturing a lost past and wish to see that past created as faithfully as possible. In one sense of modernity, this is a contemporary phenomenon. The past cannot be recreated in its totality and all new design, bar a perfect reproduction, involves design skills and a degree of invention. The desire to emulate the past is, in any case, a modern attitude to the modern world as much as the Renaissance, neo-Classicism and the

Solar House is an energy-efficient design, where the towers vent the heated air naturally.

Gothic Revival were in their time. It would be very rare and highly eccentric to also insist on the omission of electricity or efficient thermal insulation; what is desired is generally more a souvenir of something treasured than a futile attempt to turn back the clock.

In limiting the repertoire of what can be classical, however, great opportunities can be lost. In the first place, the later stages of evolution tend to be overlooked. In the 19th century in particular, the introduction of new materials offered new opportunities. Architects at that time, steeped in the classical tradition, did not seek out deliberate novelty but extended the tradition much as it had been extended in the past. Cast iron allowed structures to be made much more slender and a whole new repertoire of narrow columns was introduced. (These were not, in fact, completely novel: surviving Roman examples in Hadrian's Villa and Pompeii seem to be versions of the exaggerated fantasies of slim details in third-style wall paintings.) Improved glass manufacture and iron structures allowed for lightweight spans filled with glass and, out of this, railway stations, shopping arcades, and conservatories established a new range of classical types. Industrialization generally allowed for cheap reproduction and cast-iron balconies, railings, and even whole buildings spread classical design across the industrialized world and beyond. In the early 20th century the development of the classical tradition went further still. The skyscraper, invented in the late 19th century, reached its fully developed classical form. Sophisticated and inventive details established a unique but still distinctly classical genre most famous in Sweden but found in various manifestations throughout the world. The introduction of formal asymmetry, drastically simplified detail, stylised sculpture, and new details such as glass balustrades added another dimension to the classical tradition as significant as the new interpretations of the Renaissance.

At one level, it seems to be a needless omission to ignore this invention in the pursuit of a fictitious historical purity. Beyond this, the frame of mind that does so is also unlikely to continue the evolutionary process. Opportunities for further development abound. Superior new materials, such as stainless steel or glass-reinforced concrete, offer new expressive opportunities. Large areas of plate and toughened glass were rationalized into the classical vocabulary in the early 20th century and their possibilities are not exhausted. Developments in structural engineering allow the realisation of objectives we know to have been frustrated by available technology in the past. None of this implies the abandonment of the classical tradition any more than the incorporation of the arch into the Greek post-and-beam system, the bottle baluster in the Renaissance or the iron beam in the 18th century. Classical architecture has not been a pure expression of construction since the introduction of the pilaster or the twin-skinned dome. A tradition is not identified by a dogged adherence to its previous manifestations but by the ability to recognise its origins in its subsequent development. It requires only literacy and a commitment to a cultural tradition to maintain its continued evolution.

Classical architecture and detail should continue to evolve. Like biological evolution, both past success and simple redundancy will survive; the familiar and the canonical will persist. Like biological evolution, there will be failed and successful innovations; practitioners should not let the fear of failure drive them to stagnation. And like biological evolution, invention and adaptation will be the key to the survival of classical architecture in a changing world.

The capitals at Solar House are modeled after lotus leaves.

HENBURY HALL (see page 133): Artist Carl Lubin was commissioned to paint several views of this new country house which was inspired by Palladio's Villa Rotunda.
Julian Bicknell & Associates

The Craft of Classical Details

Julian Bicknell

古典建筑细部的工艺

ONE OF THE CHIEF DELIGHTS of working with classical details is the positive response of the craftsmen. Millworkers, plasterers, stonemasons, and the other craftspeople on a building site don't usually have the understanding or affection for traditional detailing that we expect from a designer with a classical training—or indeed an autodidact like myself with a late-blooming enthusiasm for traditional architecture and construction. And yet, in my own architectural practice, I have frequently found that craftsmen coming to classical or traditional work for the first time, react with a heart-warming enthusiasm, usually absent when faced with the so-called "modern" vocabulary of abstract forms and "negative" details.

The reasons for this reaction among building craftsmen are complex and varied. But the anecdotal evidence lends weight to the argument that traditional forms and details have some intrinsic significance beyond the purely subjective—beyond issues of style and fashion, or of theoretical discussions of perfection and the sublime. What is it in traditional and classical detailing that appeals so directly to the craftsman?

First, clearly, is the joy in making. Making things is a pleasure everyone knows from childhood—cooking and sewing, drawing and painting, woodwork and model making, covering the whole field of human creative activity right through to music and poetry. Likewise any architecture or architectural detailing that require the exercise of the physical and mental skill of making things is clearly more rewarding than the purely mechanical—as the Arts and Crafts movement of the late 19th century strove so eloquently to demonstrate.

Part of the joy of making things is in the logic of construction: the successive steps in the process of making the parts and assembling a whole according to a predetermined plan. Traditional architecture is full of details that reflect the process of construction. Take a paneled door, for instance, with its interlocking stiles and rails, the grooves and mitered moldings that frame the panels, and the fielded panels themselves; together forming a rigid and lasting whole without dependence on secondary fixings such as nails or screws. Or the elegant logic of a three-part architrave in timber for a door or window: the inner plane formed from the frame itself; the intermediate fascia forming a mask to the plaster junction; and the outer molding providing a robust edge and visual frame to the whole composition.

The inverse of this logic of assembly is the logic of carving or modeling, the development of forms from within an amorphous solid by the application of three-dimensional geometry; followed by the cutting or shaping—whether using the quasi-mechanical methods of the jig, or the free-hand carving with chisel and mallet. I recall with pleasure a morning spent with Dick Reid making a simple straight-leafed capital, starting from a solid piece of pine, first cut to a cuboid block, then turned on the lathe to make the enclosing trumpet and finally cut free-hand with mallet and chisel to form the blades and points of the individual leaves. In each case the sequence of processes has a logic that is both simple and subtle, elegant and appealing—the intellectual pleasure complementing the exercise of physical skill.

Traditional architectural forms have a hierarchy of sophistication related to the skills of the craftsman. A stonemason in the mediaeval tradition (that continues to this day) develops his skills through several stages. He starts with the forming of rectangular blocks from rough stones, mastering the formation of flat surfaces and the laying out of perfect right angles in two and three dimensions. He then progresses to the forming of moldings, at first a simple chamfer or straight-run *cavetto* on a rectangular block, and later to ogee and decorative molding on the curve, and later still to simple figurative moldings such as egg and dart. Only then does he graduate to free-form leaf-work or full-on figurative sculpture in the round. (And let it be noted that in the mediaeval tradition, only then would he advance to the laying out of whole buildings—architectural composition.) The same hierarchy of skills, and the forms that flow from them, are found in the vocabulary of classical architecture (and indeed in other traditional architectures all over the world from Egypt to Japan). The process of learning these successive stages, of mastering the increasingly refined levels of craftsmanship, is a source of pride and pleasure to the craftsman as well as to the viewer of the finished work.

Another reason for this positive response in craftsmen is the sense of contributing to the whole. No building is the work of a single hand—nor yet of a single intellect. Each craftsman is part of a team, working from his own area of specialization, contributing a significant part to a well-ordered and beautiful whole. The relationship between craftsmen (and between craftsman and designer) is itself a source of pride and pleasure—a mutual reliance and in many cases a mutual admiration. The craftsman's interest and enthusiasm is surely enhanced, if he can see that his contribution is related to other parts of the project so as to make a well-formed whole, a thing of beauty.

This architectural ordering depends on harmony in the overall composition and good judgment in the proportion of the parts. The classical system of architectural vocabulary, grammar, and syntax has evolved in the hands of successive generations of craftsmen, designers and thinkers over many centuries. A diligent student of classical architecture will quickly discover how to compose a building with the elegant composition and proportionate details of a Roman temple or an 18th-century country house. And he can be sure that the craftsmen employed on his building will respond enthusiastically to such a composition.

Classical and other traditional architectures revel in imagery. From the largest composition to the smallest detail, classical architecture continually works with visual cross-references. Details are frequently derived from a fictive structure—a fire-surround employs the columns and entablature borrowed from of a full-scale building. Details frequently employ representations of natural or man-made forms—the acanthus leaf, the interlaced vine, egg and dart, plaited rope patterns, bead and reel. Abstract forms may be derived by translation from familiar objects—balusters from pottery vases, triglyphs from timber beam-ends, the scrolls of the Ionic capital by some imagined toffee-like metamorphosis of the Doric abacus. Casual passers-by may not pick out all these connections for themselves, but they will understand them subconsciously from their own experience of plants, of carpentry, of ceramics or of toffee. And when these connections are pointed out, the reaction is one of

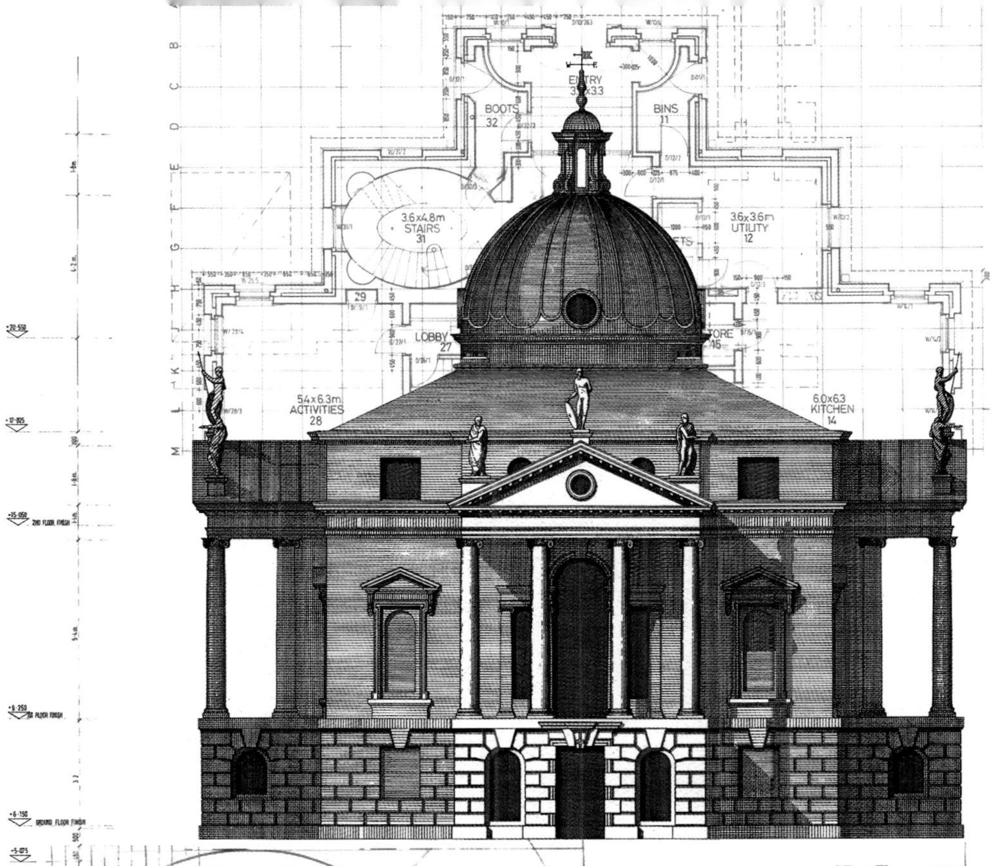
BOOTS
32
BINS
11
3.6x4.8m
STAIRS
31
3.6x3.6m
UTILITY
12
29
LOBBY
27
5.4x6.3m
ACTIVITIES
28
6.0x6.3
KITCHEN
14
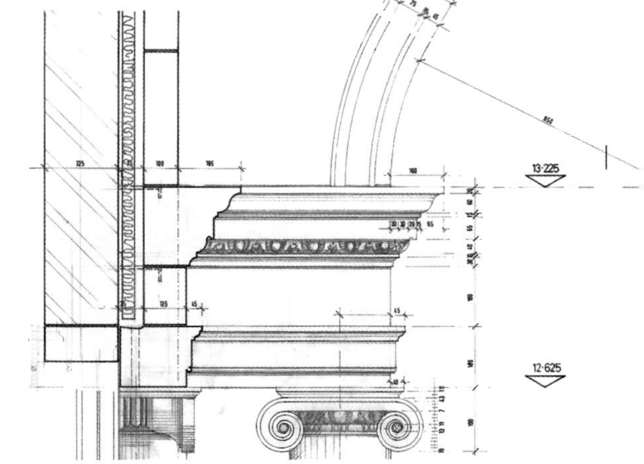
13·225
12·625
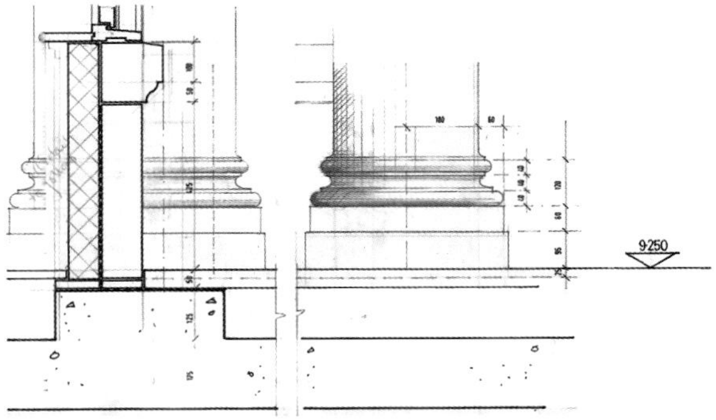
9·250
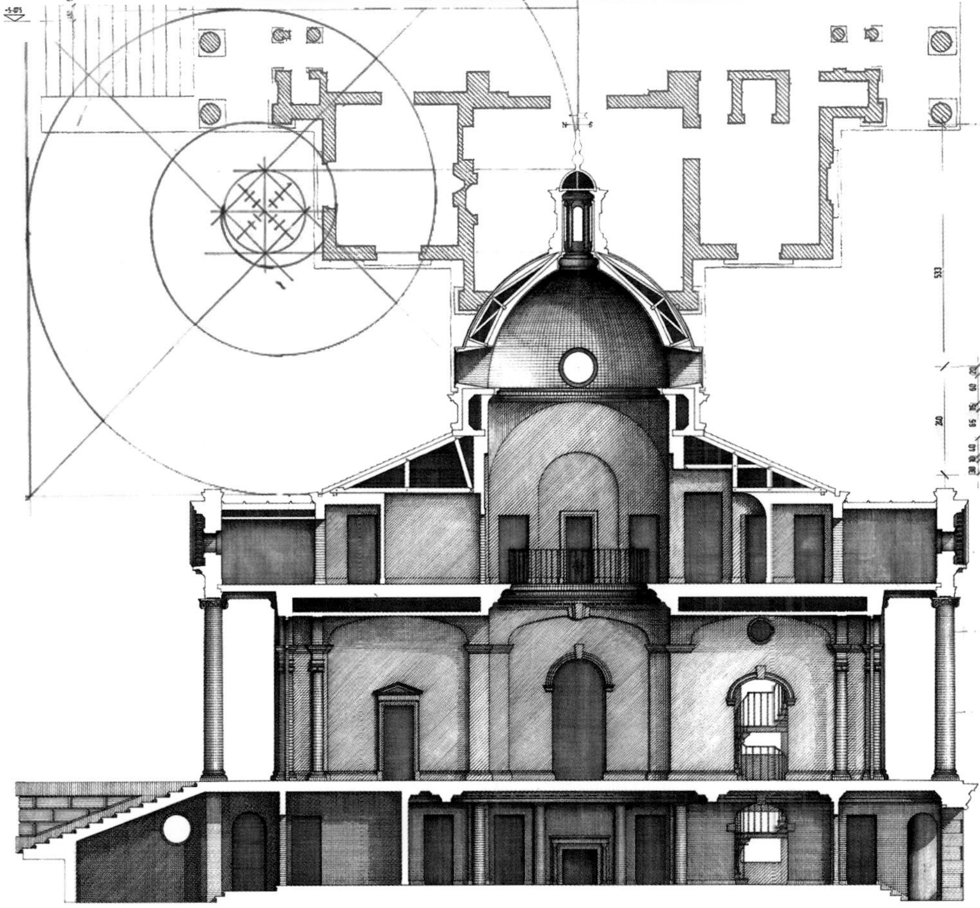
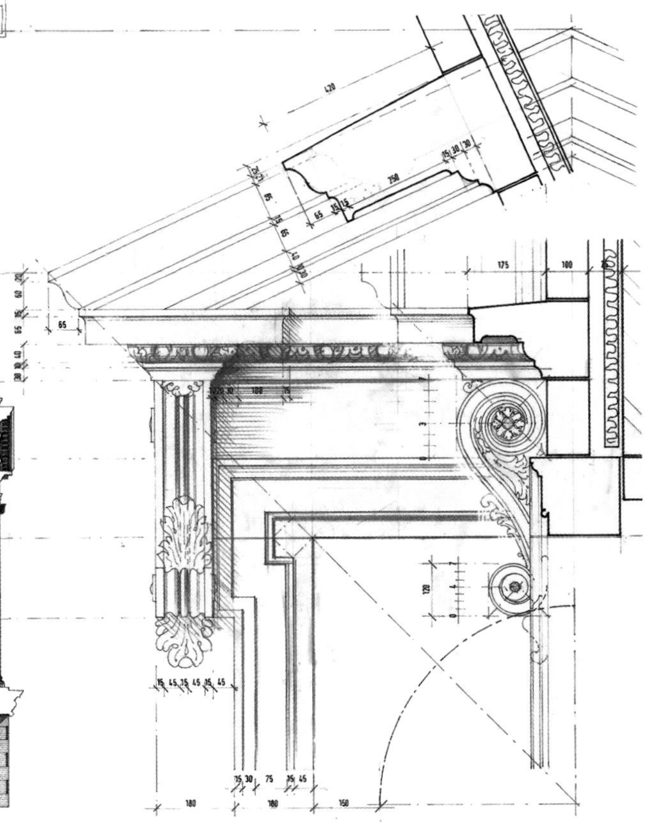

delight and recognition rather than the bafflement evoked by so much "cutting-edge" modernism.

Our craftsmen also react very positively to originality. The fact that the classical tradition is so well established does not preclude an element of creativity—even of fantasy. The history of classical architecture is punctuated in every generation with examples of striking innovation in both form and detail: the multi-lobed vaults of Hadrian's Villa; the idiosyncratic details of Michelangelo's Medici Chapel; the extravagant multiplicity of cornicing in the Sacristy of the Charterhouse at Granada; the structural gymnastics of Guarini's St. Lorenzo in Turin; the curvilinear plan-form and studied syncopation of Neumann's Vierzehnheiligen; the minimalist elegance of John Soane's Dulwich Mausoleum and Picture Gallery. Each of these examples, and many others like them, stretched the skill and understanding of the craftsmen of the day beyond that which even the most experienced will have undertaken previously—a challenge which I am sure they relished. And where the result of their efforts is a source of wonder, amazement and delight, the craftsmen share the pride and pleasure of creation. The whole is much more than the sum of its constituent parts.

It is possible to enumerate and analyze many other aspects of the world of classical details. But the designer is often unconscious of the sources of his inspiration and the craftsman probably uninterested. The key point is that both the designer and the craftsman are working in a world of understanding so well as established as to become intuitive. The vast encyclopedia of traditional forms and details is a constructional and visual language that each in their different way has learned, that each comprehends; and each can use as we do our native spoken language—without conscious thought for the intricacies of syntax, grammar and vocabulary.

And further, the man-in-the-street who then lights on the finished work also becomes an heir to the tradition. Again, without thought, he can respond to the forms, and recognize—even if he does not fully understand—that this is a language, deriving from the essentially mundane business of designing, making and decorating elementary shelters, that has a wealth of depth and meaning that makes it one of the most enduring forms of high art created by mankind.

PAGE 71:
An analytique of Henbury Hall, where Julian Bicknell shows plans, elevations, sections, and details, all on one drawing.

OPPOSITE:
Constructed of yellow cedar, this 1:50 scale model of Henbury Hall, was used to test critical aspects of the design—in particular the use of a fully detailed cornice round the main block, unlike Palladio's Villa Rotunda where the cornice is reduced to a decorative band course. Built by Richard Armiger, the model has been exhibited several times, including at the Victoria and Albert Museum in London.

ASHFOLD HOUSE (see page 209): Conceived as a triumphal arch, the front door is contained within a central vaulted archway recessed into the façade.
John Simpson & Partners

Building to Last

John Simpson

屹立不倒的建筑

IT IS WORTH PAUSING to consider how we, in the 21st century, have come to have such a rich language of Classical Architecture and such a treasure trove of architectural details to admire, adopt, and develop. In the old city of Split in Croatia, the two vital architectural eras that shaped Western architecture are intermingled: Ancient Roman walls and arches weave amongst Renaissance buildings. The strong Roman walls survived the break-up of the Roman Empire and because they were still standing by the fifteenth century, they, and other Roman buildings across Europe, inspired the architects and artists of the Renaissance. The revival of classical design and its application and adaptation to create new types of buildings since the Renaissance has endowed us with a rich legacy of buildings and public spaces. Key to this continuity and development of the classical tradition is the durability of the construction of the buildings. Yet why did our predecessors build to last? Focussing on the Ancient Greeks, one can see how important architecture was to their sense of identity and how the durability of their buildings was intended as a statement of the longevity of their own existence. The details of their architecture proclaimed the Greek identity wherever they were built. However, intervening centuries have shown that durability lies as much in a building's ability to be adapted to new uses as in its materials and method of construction.

Architecture, above all, is about identity. Whilst identity is not static, it relies on the continuity of fundamental identifying features. This is true also of the architecture that expresses that identity. The Ancient Greeks began by building in timber but as they became more prosperous, timber was superseded by stone. Yet despite this fundamental change in construction material, the Greeks did not adopt the use of the arch. The Greeks would have known of the ancient examples that can still be found around Memphis, Western Thebes, and Babylon and, from these, the fact that the arch is a cheaper and easier form of masonry construction. Yet they stuck solidly to the post and lintel method of construction and the concomitant architectural details. The arch was not part of their heritage but post and lintel construction, reminding them of their origins, would have been seen by them as definitively Greek. Therefore every building constructed in this way reaffirmed their values, their ideals and their traditions. This architecture united them and set them apart from the surrounding barbarians. It gave them their identity. Moreover, it stamped their identity on the territory they controlled with the solid masonry giving a clear message that the Greeks intended to remain in control forever. By making their architecture durable, they ensured the survival of their identity and their culture as a society.

Via the Renaissance, the West has inherited a language of architecture from the ancient Greeks and Romans that we use and develop up to this day alongside our spoken and written languages to reflect and to constitute our sense of identity. Inscribed into the buildings that form public spaces, the architectural language embodies the values that bind our society together. Our values, history and shared memories are intertwined with the civic spaces and their architecture. Take, for example, Trafalgar Square in

London, which, like all traditional squares, is made up for various individual buildings that define its perimeter. The space within is a public space and is the site of, among other things, public gatherings for New Year's Eve, the national Christmas tree, public celebrations and political demonstrations. Trafalgar Square has become a principal national public space because it has hardly altered since the mid-19th century. Buildings that last give a sense of permanence to our identity and an impression that the values intrinsic to that identity will endure. So today, just as it was in the ancient past, durability lies at the very heart of what architecture is all about.

Logically a building that is meant to last needs to be built from durable materials arranged in a stable structural form. This principle suggests that the Ancient Egyptian pyramid, a supremely stable form built from solid stone with a granite facing and without structural weaknesses such as doors and windows, would be the ultimate exemplar. Yet the pyramids are a very poor model for buildings designed to last. Like other structures from the ancient world, they have been quarried for building materials because they served no other useful purpose to the successive generations. Two structures built by the emperor Hadrian illustrate the importance of continuous use to the survival of a building: whilst his villa at Tivoli now stands as a pillaged ruin, the Pantheon in Rome has remained in continuous use since the 2nd century AD and consequently stands intact. Similarly, the timber Shinto temples in Japan, which are exactly rebuilt every 25 years by a particular order of monks, suggests that it is as important for a building to have a purpose than to be built of durable materials.

In the 21st century we need a practical workable model that combines durability of material and construction with the principle of the importance of the building's usefulness and flexibility to change. Such a model can be found in the buildings in Bedford Square in London, built by Thomas Leverton between 1770 and 1780. The buildings have retained their elegant façades and the square has consequently retained its character. However, over the centuries the buildings have been used for a variety of purposes: homes, offices, hotels and even a school of architecture. Whilst Bedford Square has the advantages of being well designed, well built of durable materials and having weathered well, its significance as a model lies in the hierarchy of construction that each building represents. The internal partitions are built of materials such as timber that can easily be rearranged to accommodate different uses whilst the façade of the building is elegantly designed and well built from durable materials, such as brick and Coade stone. The exterior of the building and the building structure, the parts of the building that belong to and serve the public realm, are built as durably as possible whereas the parts that serve a private use have a lesser degree of durability that enables them to be altered to serve the changing needs of each successive generation.

Employing a hierarchy of construction is vital to the longevity of a building. Incorporating future flexibility through the internal use of materials that can easily be removed means that the same building can be refashioned to serve different generations in different ways. Simultaneously the durability of the external façade ensures that the contribution the building makes to the public realm endures. In this

Designed by John Simpson & Partners, the penthouse apartment at Carhart Mansion includes a full Doric pavilion that allows access to the rooftop terraces.

way it contributes to the local character of the place and to a sense of identity. The connection made by the ancient civilisations of Greece and Rome between identity and durability of materials and construction still holds true in the 21st century. What we have learnt from them, however, is the importance of buildings being able to adapt to the requirements of successive generations. The details, which distinguish the architecture and resonate with identity, can survive by being adopted and adapted in new buildings.

The Arabesque in Classical Architecture

Peter Pennoyer

古典建筑中的阿拉伯式花纹

THE STORY of the great American architectural practices of the first half of the 20th century requires an understanding of the Beaux-Arts School, where many of the founding members of the profession were educated. They learned academic subjects such as mathematics, geometry, and history, while steeped in the transformative and intense experience of the atelier. These studios were the crucible of the creative process—places where apprenticeship meant learning by drawing—on others' projects as well as one's own.

The academic framework of the Beaux-Arts School was built around the theories of architecture represented by various interpretations by great French thinkers from Roland Fréart de Chambray to Claude Perrault of the classical canon. The treatises these men wrote and illustrated in the 17th century were the beginning of the dominance, by France, as the leader in assessing the complex history of the classical orders. These assessments became, in turn, the basis for theories of practice. The tension between the idea of the primacy of the orders and the theory that structure and rational design process drove architectural design presaged the battle between the modernists and traditionalists that has never completely ceased. These positions color our views of most significant buildings.

An astute observer will understand, intellectually, how a building fits into the story of evolution of classicism. This sequence is not hard to trace: from the discovery of the manuscript of Vitruvius's *De Architectura* in Saint Gall in 1414, Alberti to Serlio and those who followed have engaged in a dialog, spanning the modern era, about how to make architecture. Each has had some measure of influence on practice. The lineage can be illustrated by comparing plates from treatises to examples of built work.

Closer investigation sometimes leads right to the source of an architect's inspiration: interior cornices in Monticello appear to follow the proportions shown in Claude Perrault's 1668 *Vitruve*, a treatise that Jefferson owned. This kind of academic detective work, neatly packaged in courses and histories, naturally focuses on the pure examples of uncluttered styles, creating a crystalline comparison. What is missing, naturally, are the more complex and obscure elements within each period. Among these aspects, none is more challenging to pin down, in terms of its origins, than the Arabesque.

The Arabesque is a curvilinear, flowing, vegetal decorative pattern that most commonly appears within panels and in three-dimensional decorative elements such as bas-relief and in architectural metalwork. Perhaps the obscurity of its origins is the reason why it infuses classical architecture with such variety, interest, and depth.

In the work of McKim, Mead & White, we see how a synthesis of both Islamic and Roman Classical sources testifies to the ability of Stanford White as well as his Beaux-Arts trained colleagues, and the artists with whom they worked to transcend direct stylistic quotation and present hybrid forms of stunning originality. Unlike the carefully controlled system of proportions that govern the classical orders, the Arabesque defies overarching rules. While suggesting complex geometries, it does not follow prescribed geometric formulae. It may suggest the great mathematics of Islam, but its forms are not explicitly based on them.

In the Villard Houses, the firm's great Italian Renaissance masterpiece in New York City (1885), McKim, Mead & White designed something unprecedented, infusing classicism with artistic force, creating a total work of art that had never been seen before in America. They stretched the limits of the classical canon, transforming historical precedent into a fully realized aesthetic synthesis.

The architects wove a complete architectural cloth from disparate styles and elements. For example, in the front hall and stair of Henry Villard's mansion—one of six in the complex—the delicacy that infused Islamic detailing that we see in some of their earlier works, particularly their Shingle Style houses, is brought to this house, giving the interiors a lighter touch—what we might call synthetic eclecticism—from the mosaic vaults and floors to the inlaid marble walls. In the tympanum, attributed to Augustus St. Gaudens, and in the mosaics, attributed to glassmaker David Maitland Armstrong, we see the notions of geometry from the full range of influences of the ancient world.

Italian Renaissance details in the dining room are infused with the delicacy of Islamic arabesques. Figures of Joy, Hospitality, and Moderation by St. Gaudens in the mantelpiece and leaping fish sculpted in the niches enhance the room's artistic dimensionality. Intricate tracery of stylized leaves and flowers on the dining room ceiling draw inspiration from Moorish details of the carved ivory caskets of Caliphal Spain, where floral arabesque patterns merged with bird, animal, and human images.

The main stair and the St. Gaudens zodiac clock reflect further inventive stylistic syntheses: Stanford White and St. Gaudens collaborated on the zodiac wall clock on the landing, and John LaFarge created the Renaissance putti, arabesques, grotesques, and Pompeian details in the window. Finally, the music room included the mural "Drama" by John LaFarge with sculpted ornament, again by St. Gaudens, plaster casts of Luca della Robbia's marble Cantoria from the left sacristy of the Cathedral of Florence, and a window of overwhelming scale by John LaFarge. The sheer artistic unity in this space is especially notable given the range of artistic temperaments and materials brought to the commission.

From the beaux-arts to modern day classicism, we can again see how fluidly Islamic geometry permeates classical details, elevating the tone of the design to a sublime level of mysterious delicacy.

PAGE 78:
DRUMLIN HALL (see page 189): Faux marble columns and a hand-carved anthemion frieze mark the entrance to the library.
Peter Pennoyer Architects

OPPOSITE:
Designs for a wrought iron balustrade and newel post, featuring stylized leaves and flowers
Peter Pennoyer Architects

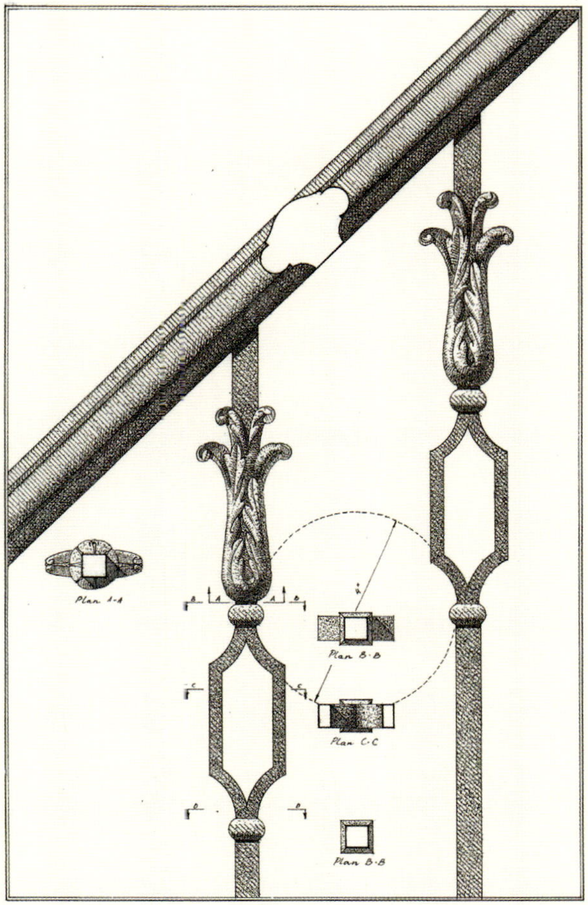

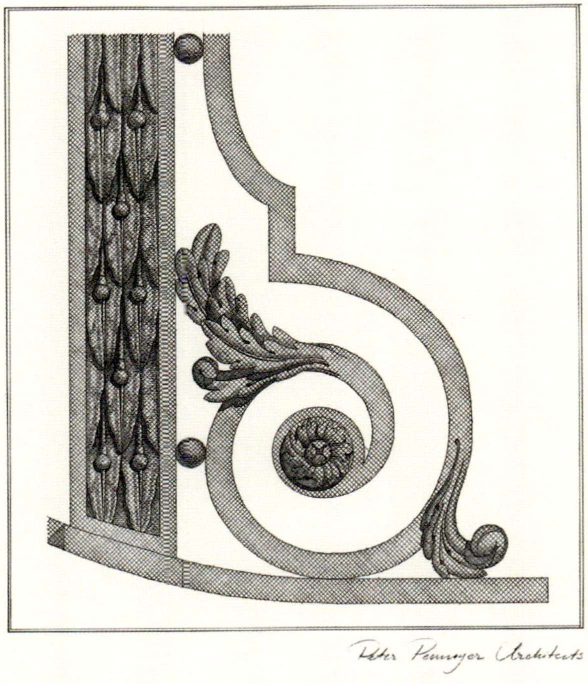

Peter Pennoyer Architects

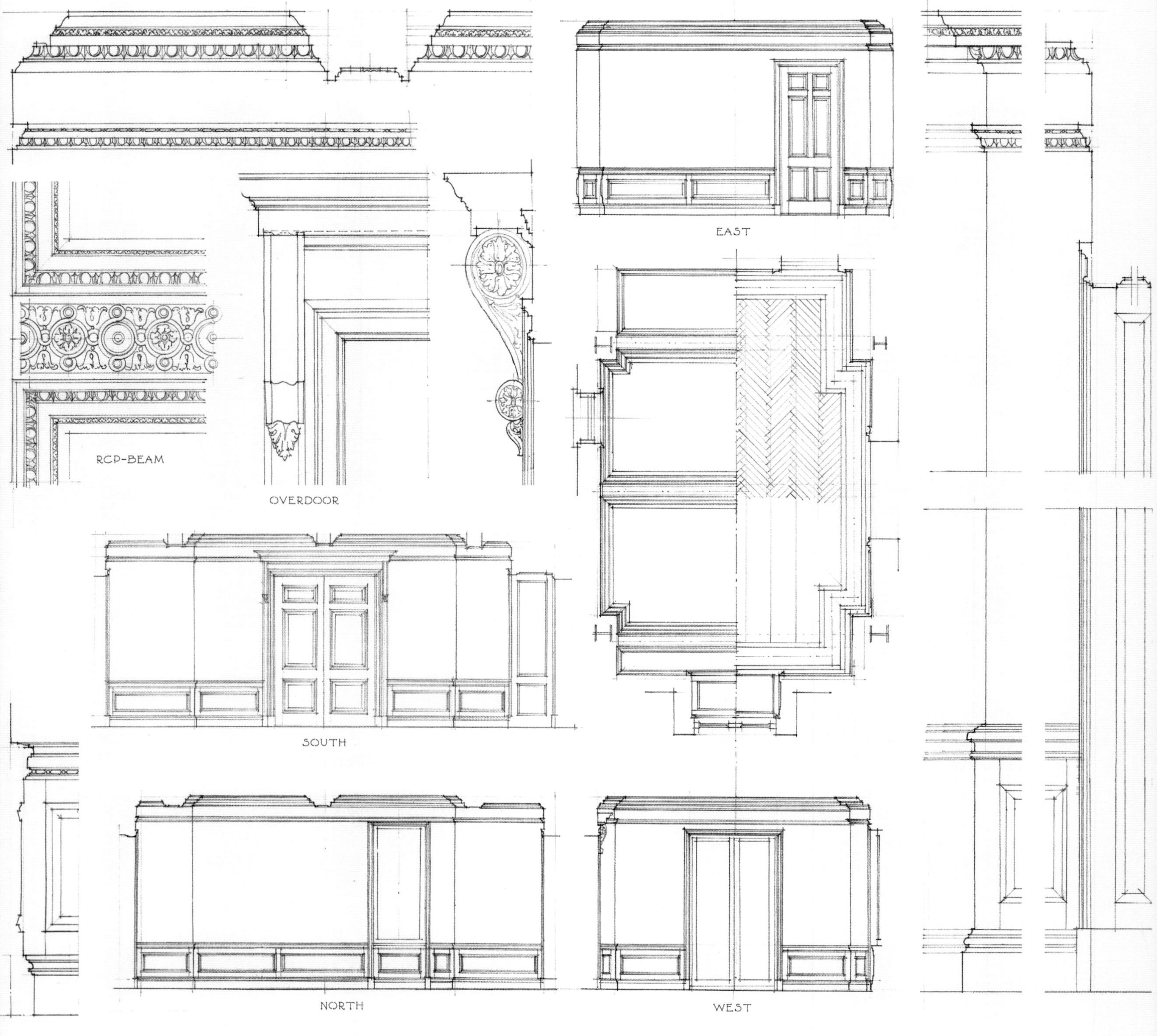

John B. Murray Architect, LLC
48 West 37th Street, 10th Floor
New York, New York 10018

Design Study for a Residence

New York, New York

Dining Room

Design Development and the Analytique

John B. Murray

设计开发与分析

AT A TIME when stunningly innovative software that seemingly brings the design process cutting-edge efficiency is available, it might be perceived that keeping alive the methodical craftsmanship of the Beaux-Arts School is "quaint." Yet when I began as an architect, more than 30 years ago, I was intrigued and inspired by the Beaux-Arts commitment to classically high standards that insisted upon "a compositional analysis, an analytique," design concept for any building, façade or interior. Centuries later, using the template of analytique drawings as multi-visionary tools for producing the highest of quality results called to me as still very modern. To this day I am impressed by the joy and creative excitement the analytique design process brings to my clients. To create a space of one's own, whether from the ground up, or by working from a seemingly mundane shell, must be a well-considered journey of proportion, light, space, texture, depth, interior, and exterior views. Through my years of collaboration with clients, designers, masons, specialists in lighting, heating, and climate control, the craftsmen still creating plaster cornices and intricately laid mosaics of stone or perfectly set variations of wood, I have continued to appreciate, yes, of course, the refined art, in and of itself, of an analytique rendering, but more importantly its ability to communicate design details in a manner that permits all involved to envision the importance of every detail within a broader composition.

Writings on the possibilities of an analytique drawing, such as Lloyd Warren's 1922 foreword to John F. Harbeson's *The Study of Architectural Design,* continue to inspire the work of my firm. Warren describes society's love of "the short-cut way," but advocates for a "most sincere ... admiration for the excellent methodical way ... of a good *analytique rendu.*" His assessment that "in these drawings can be seen sensibility to proportion, feeling for composition, character in drawing, appreciation of ornament, and knowledge of descriptive geometry in projections and in shades and shadows" begins to illuminate the multifaceted means of communication these drawings provide. For me, what continues to make analytique drawings so successful, is as Warren once wrote, that because the drawings are produced as a result of dialogue between client and architect "everything seems as clear as day." After my team meets with the client, we illustrate design conversations with hand-wrought drawings that stimulate the client to increasingly clarify—again in Warren's words, "the personality which is the basis of all style." There is a classicism to our architectural process that is timeless. Working with our clients through drawings that capture dimensions and spatial relations, we pull out the finest of design details, creating an end product that is uniquely personal.

Our design process at 825 Fifth Avenue is a case in point. A couple from Boston approached us with the concept of creating a New York City pied-à-terre in a pre-war building on Fifth Avenue. We began design development from the known parameters of two rather nondescript one-bedroom apartments. Our clients expressed their intention to carve a "jewel" from this completely demolished space. For their urban residence they sought highly articulated, richly detailed, beautiful materials, and expert workmanship.

They understood that through the process they would be better able to articulate more precisely how the space might look and feel and how the many facets of design would fashion a "jewel" unlike any other.

Where did our analytique renderings come into this design development? From our very first meetings we presented analytique drawings with a "sensibility to proportion." We organized the circulation of the entry, visually showing the owners that space would be channeled from a rotunda entry vestibule to a rectangular gallery as well as along a vaulted hallway connecting to the master bedroom and guest room suites. Analytique drawings provided a "feeling for composition" as they visually articulated a concept of proportioning the living room to simulate an unexpectedly grand salon.

In this project, our hand-wrought drawings were most helpful in the design process as artistic dialogue to uniquely express client "appreciation of ornament." Through drawings we could show how highly stylized cornices with a classic leaf pattern would look in the bigger scheme of the living room. Other drawings explained that the fireplace, with custom bronze work cast by P. E. Guerin, would be a unique focal point in itself, but would also be integrated into the paneling of the south wall of the living room. In another design presentation we illustrated how a coved ceiling with applied molding would serve as a decorative organizer for ceiling lighting. Still another drawing illustrated how a pair of Harmon hinge doors connecting the living room to the master suite would facilitate closets within the wall thickness. Each analytique was an opportunity to show the big picture of a room along with an "appreciative" close-up of the ornamental details.

The analytique drawing represents multiple views and permits clients to readily perceive the concept and the cohesiveness of the design. The analytique forces us as architects and designers of living space or form to depict all the components of a project in one drawing. The analytique drawing quickly reveals the quality we will provide. Through meticulous drawing we send the message to client and builder that this is the high standard we will maintain through completion of the project.

In the design process problems that arise trigger the design and the design triggers the solution. Since drawing scale from a variety of perspectives is critical in an analytique rendering, resolutions to design questions present themselves as the architect, designer, and client view every corner, every surface, the minute, and the grand view. The architectural design process must be precise in scale and proportion, but open to the creative synergy of the process itself. Analytiques are teaching tools that enable all involved in the design process to conceptualize space, to visually embrace how all the components fit together to honor a classically masterful whole. At their best analytique drawings are classical art. Project after project, I have seen their effectiveness in design development. It is not unusual for a client to request our analytique be framed as a memento of the pleasure of the design process.

No, analytique renderings are not design development "short cuts," but they are an old-world art form that continues to shine a focal light on the brilliant magical promise of ornamental detail in service to a finely executed whole.

FIFTH AVENUE PIED-A-TERRE (see page 185): The living room is enriched with wall panels and an intricately carved pulvinated frieze.
John B. Murray Architect

A rendering in chalk by Richard Cameron for a portico at a private residence

Drawing in Full

Richard Cameron

完整绘制

"*PRACTICE BY DRAWING things large, as if equal in representation and reality. In small drawings every large weakness is easily hidden; in the large the smallest weakness is easily seen.*" L. B. Alberti

Drawing is so fundamental to design that our most basic attempts to explain something usually involve picking up a pen or pencil to "diagram" an idea. Architects sketch incessantly—even in the age of computer-aided drafting and modeling. Yet more and more the development of design ideas is left to the digital realm so that after the first schematic explorations, drawing is abandoned for the scale-less, characterless world of software programs and flat screens. This seems like a profound mistake to me, not least because, as Alberti says, the large-scale drawings that were typically part of the later stages of design development, most reveal faults in our designs—and consequently are the most worthy of sustained study.

Traditionally, design begins with small-scale sketches and studies of various ideas. These sketches are used to explore alternative design approaches and to establish the most general appearance and function of a building. They are often carried out in soft pencil or charcoal and are sufficiently vague that the formal and functional aspects of the design remain fluid and flexible. The basis for these sketches lies in the designer's imagination, but that imagination is filled with the experience of the work of other architects, of other designs, and it is the influence of that other work, both conscious and unconscious, that pushes the design in one direction or another. In the modern era the attempt has been made to reject this influence with results that we see all around us. For the traditional architect however, the influence of the work of past masters is embraced and there is a self-conscious effort to adapt previously successful design solutions to present design problems. Thus the search for a "parti" for a design problem is also the search for a model upon which to base a design, chosen for its appropriateness and applicability to the challenge at hand. In the Beaux-Arts teaching system, the discipline of establishing the "parti" was honed over years of practice in student competitions (which formed the criteria for advancement in the school). By the time an architect had graduated from this system he or she was highly efficient at coming up with the basis for a design solution to any given program.

Beyond the ability to solve a given design problem efficiently the Beaux-Arts approach taught its students to stick with the "parti" and to systematically develop it into a highly refined piece of architectural design. The great strength of this system lay in this process of refining a fixed design idea. It equipped the architect to face complex issues and make the most out of the contingencies that one faces in real world design problems.

In the early years of the system, young architects were taught to compose their design solutions into drawings known as analytiques. These were compositions that attempted to represent all of the scales of a given design from overall plans, sections, and elevations to large-scale details in a single drawing. While these were rarely full-size details, they were nevertheless depicted at scales large enough that almost all design

RAVENWOOD (see page 141): The rear façade of the house recalls English architecture of the 17th and 18th centuries. *Richard Cameron**

refinement would be evident. Through years of this practice, the typical Beaux-Arts trained architect would systematically refine his design through a range of scale drawings, each of which would be meticulously represented in the exacting method of India ink wash.

The Atelier system that this took place in was arranged to resemble an office. Each Atelier was lead by a senior architect and the students were in effect apprenticed to this figure. Older, more experienced students were responsible for much of the teaching of basic skills, and in turn the younger students helped the older on their competition entries, gaining valuable experience. Many of the Ateliers led directly to the actual offices of the Atelier heads where work was carried on in the same spirit. Thus the student moved from the world of school to the world of practice in a clear and systematic progression. Most of the famous offices of the early 20th century worked in this way and the consistently high standard of work they achieved is a testament to the strength of the system. In particular, the discipline of rapid and efficient design solutions followed by sustained refinement in detail up to and including the many full-scale details required for a traditional building was the principle method of design practice carried over from the schools. The beauty of Beaux-Arts buildings lies not only in their clear practical and formal resolution of the program and circumstances of the design, but in the exquisitely refined details of each building and in the way these details relate to the whole composition.

A superb example of this achievement is the New York Public Library by Carere and Hastings. Both men were products of the Ecole in Paris and were alumni of the office of McKim, Mead & White. In fact, they beat their former office in the competition to design the library. This building on Fifth Avenue is unarguably one of the finest buildings ever erected in this country and it continues to function today exactly as it was intended 100 years ago. It has recently received a major restoration and so once again, the splendid white marble of the building reveals the superb artistry of its designers and craftsmen. What is most striking about the building is that at whatever scale it is examined, from the largest entablature to the smallest drinking fountain, it is evident that its architects controlled and refined every aspect of the design through large scale drawings and models. Alberti's concept of "concinnitas" comes immediately to mind such that it is possible to assert that nothing could be added or taken away from this building that would improve its design.

Alberti's admonition to draw at full scale is one we should once again take to heart. The joy of great buildings is the way in which they speak to us and move us. As humans we delight in everything from ornament to order. In drawing and modeling the things we design at full scale, we are compelled to struggle with their weaknesses large and small. Fortunately for us, we have masterpieces like the New York Public Library close at hand to show us the way.

A rendering by Richard Cameron of the garden portico at Ravenwood

The entrance façade of this new Regency style villa in Cheltenham includes flanking walls that define the boundaries of the entrance court.
Hugh Petter, ADAM Architecture

The Future of the Crafts

Hugh Petter

工艺的未来

"....of course you could not do that nowadays because, the trouble is, there just aren't the people with those sorts of skills anymore."

HOW OFTEN HAVE WE HEARD this urban myth trotted out as we look at some highly crafted historic object? Anyone could be forgiven for thinking that the crafts in the early 21st century are in a state of terminal decline. Myths, of course, develop for a reason and so it is worth looking for a moment at some of the issues that may underpin this widely held belief.

Increasingly onerous health and safety legislation and employment law over the past 30 years have made the prospect of running a large craft studio an onerous one which many craftsmen have chosen to avoid by working in much smaller groups. As a consequence, the majority of craft-based businesses employ less than half a dozen people and are often based in sheds at the bottom of gardens, or in converted industrial buildings with little kerb-presence. These businesses, therefore, often have a very low profile and rely upon the support of local customers and well-informed clients who are seeking a specialist service. Their relatively obscurity certainly adds fuel to the myth as people struggle to know where to go to find these specialised skills.

The rise of the conservation movement after World War II certainly sustained a number of traditional craft skills. Formed initially to repair war-damaged buildings, and as an antidote to the brave new world of International Modernism, the conservation movement has provided plentiful work for generations of craftsmen, albeit on the repair of historic buildings rather than in originating new work. This predominance of repair work to fragile historic structures often requires old-established techniques to be employed in preference to more modern technology: in turn, this can create an impression in the minds of some that traditional crafts skills are stuck in a time-warp and are therefore increasingly irrelevant to the challenges facing more mainstream new building projects.

People who want to study a traditional craft may struggle to find information about all of the courses which are available because many of the institutions where these subjects are offered—for example The Building Crafts College or the City and Guilds College in London or the American College of the Building Arts in Charleston—are independently owned. These colleges offer courses of very high quality but they tend to operate in relative isolation and so their profile is rather lower than the larger, more mainstream higher education institutions. This relative obscurity can make it difficult for some students to find the courses they are seeking, so fuelling the myth that it is no longer possible to study traditional building crafts.

In recent years, governments in the Western world have been keen to promote craft training as a viable alternative to an academic university qualification. Significant sums of money have been available for apprenticeships. However, this is not always as well targeted as it might be—there is money available to support students who want to study craft courses from many different sources, but it is often so poorly coordinated that students struggle to find it. Similarly, upon graduation, craft students benefit from spending some time with an established practitioner to acquire professional experience before launching their own business. However, because of the nature of the majority of craft firms, few established craftsmen are willing to saddle themselves with an apprentice as this brings with it not only a loss of productivity for the business, but a whole raft of employer responsibilities which they would rather avoid. As a consequence, many graduates from crafts courses end up pursuing careers in other fields. This too adds to the impression of a system that is in terminal decline.

Yet all is not gloom and doom. Indeed, green shoots abound: for example, The Art Workers Guild in London, founded in 1883, was the cradle of the Arts and Crafts movement which quickly spread all over Europe and across the Atlantic to the US. It is a club where new members (called Brothers regardless of their gender) are elected by peer group review and where regular lectures and events are arranged to provide a point of focus for the members, many of which otherwise work in relative isolation. It is currently growing at a rate of about 15 percent a year and its 300 international members span some 80 craft, fine and applied arts disciplines. Over the past 130 years The Guild has spawned over 80 other kindred societies, many of whom use its building in Central London for their meetings. It is estimated that over 30,000 people who are connected with the decorative, fine and applied arts cross its threshold each year and yet it has never received a bean of public money. The Building Crafts College in London has over 1,000 students and is expanding in response to the demand for the courses that it offers; the American College of the Building Arts in Charleston is similarly in good heart and has grown very quickly from its foundation ten years or so ago to become recognised internationally as a significant institution. In Russia, the museum at Tsarskoe Selo near St Petersburg set up a craft workshop to rebuild the Catherine Palace, which had been destroyed during the German occupation. That workshop now employs over 70 people and undertakes both conservation work and new projects all over the country. There are endless other examples.

Whilst conservation work may well require old techniques, modern technology can help in many other areas. The internet, for example, can be used to help raise the profile and accessibility of small craft businesses and can make it easier too for people to work together regardless of where they are in the world. This in turn creates a larger and more readily accessible market for specialised crafts businesses.

Similarly, the internet is the perfect environment in which small colleges offering craft-based courses and sources of funding for students and apprentices can be presented in an accessible and organised way. The Art Workers Guild pioneered this with a dedicated website launched in 2007.

The so-called “five axis machine” was developed originally in Italy for the car industry but has found a new use in stone workshops. An object, such as a swag or column capital, is modelled in clay by a craftsman; it is then scanned to create a virtual three-dimensional computer model. The machine then proceeds to carve the stone using progressively finer chisels as work proceeds. It can carve in eight hours what

it would take a man to carve by hand in eight days. The craftsman still creates the original object in clay and finishes the stone copy by hand, but much of the relatively mechanical, time-consuming and mundane work of carving the object out of the block of stone can be done more quickly and cheaply than would otherwise be the case, keeping costs to a commercially viable level and leaving the craftsmen with more time to concentrate on areas of work where their specialist skills add most value.

The writings of Pugin and Ruskin were hugely influential internationally and underpinned both the Gothic Revival and the Arts and Crafts Movement. They saw architecture as the "mother of the arts" and architects were encouraged to let the creative genius of the craftsmen shine through by allowing them to design the fine detail of buildings upon which they were working. Ruskin, ever an idealist, became increasingly frustrated by the lack of success with this aspiration, but he failed to understand that many of the craftsmen at that time had been trained to produce high-quality work designed by others, not to design it for themselves. In other words, they simply did not have the skills and education to innovate their own design work.

In contrast, in our post-industrial world, people in the Western world have more wealth, education, and leisure than ever before. As a consequence, not only are more able school-leavers going into better run crafts courses than ever before, but also people are changing jobs in mid-career and many are entering the crafts having had a first career in another subject. As a consequence, the gene-pool of craftsmen is significantly broader than in previous ages with more well-educated, articulate, and entrepreneurial people and, whilst there will always be an important place for conservation skills, modern technology in the right hands can help with innovation, connection between people and the achievement of high-quality results on new building which are realised in a commercial environment.

Philip Dodd has succeeded in assembling a strong field of contributors to this excellent new book. The essays that follow by Alain Olivier, Richard Carbino, Aidan Mortimer, Paul Chesney, and Foster Reeve, each a noted specialist in their respective field, provide collectively a fascinating insight both into the current state of the crafts and the rich roots of the traditions too upon which the current generation of practitioners are firmly grafted.

It is a time for optimism and for careful action too to ensure that the green shoots which are emerging are carefully nurtured so that they mature into strong plants which can, in turn, embellish both current and future generations of new classical buildings with fine, innovative, and vigorous details of the highest quality.

PAGE 93:
A detail of an entrance portico, with Ionic columns carved from Bath stone
Hugh Petter, ADAM Architecture

OPPOSITE:
A NEW COLONIAL HOUSE (see page 195): This design rendering by Chris Draper shows that the simplified Doric columns on the two-story loggia have been paired together to allow for a wider spacing of the bays, and unobstructed views of the Caribbean.
Hugh Petter, ADAM Architecture

Designed by David Easton, this library includes a bespoke mantelpiece, wall paneling, bookcases, and fluted Ionic columns—all fabricated by *SYMM* out of English Oak.

The Importance of Joinery in the Classical House

Aidan Mortimer

古典住宅中的细木工艺的重要性

THE CLASSICAL STYLE seems to be as popular as ever in both Britain and America; indeed its appeal is universal with classically inspired mansions presently under construction in every corner of the globe from Beijing to Moscow and Kuwait to the Bahamas. They all share a common heritage with their roots deep in 18th-century Georgian Britain, which is widely acknowledged as the high watermark of Classical architecture.

To understand the role of joinery, or millwork as it is often referred to in America, in the Classical house of today, it is well worth exploring the historical development of the 18th-century house and the importance of joinery in houses of this period.

In Britain, the catchall term "Georgian" broadly covers the period from 1714, the date George I came to the throne, through to 1830 when George IV died. Of course, the style is not specifically tied to the reigns of these monarchs and its origins in Britain lie a century earlier in the works of, amongst others Inigo Jones, and after 1830 the style continued to refine and develop in all sorts of interesting ways throughout the 19th century.

Georgian style can be broadly broken down into three stylistic periods: Palladian (c1714–60), Adam or neo-Classical (c1760–90) and Regency (c1790–1830). In America, the Colonial style echoed the first two periods and the later Federal style broadly mirrored the Regency style. In many ways the 18th-century American house was markedly similar to its British counterpart, although ideas sometimes took a while to reach those provinces distant from Britain. The materials out of which British and American houses were built during this period did, however, differ. Wood was, for example, far more widely used on the east coast of America than stone, which was rarely found locally. By contrast, in Britain there was a wealth of easily sourced building material from sandstone and slate in the north of the country to limestone, flint, and locally fired clay bricks and tiles in the south. As a result many of the most significant country houses were, where their owners could afford to flaunt their wealth, built in stone and most of the external ornament on these buildings was also in stone. On the east coast of America houses of the same period were almost entirely constructed in timber, with wood sidings on timber framed walls under shingled roofs and with porticos, colonnades, and ornate doorcases all made out of wood.

Despite these differences in construction across the Atlantic buildings did, and still do, share many common millwork details, most notably windows, doors, paneling, and staircases. Windows are usually the first element of a Classical house you notice; they give a façade balance and create a rhythm. The window synonymous with the Georgian era is the double hung sash which was adopted almost universally in Britain and America, and also Holland, although its use did not spread much further. Windows of the Palladian and early Federal period were rarely made to standard sizes or pane dimensions, but were characteristically six over six or eight over eight panes. As the 18th century progressed glazing gradually increased in size and four over four and even two over two panes became commonplace

Window frames in Britain were generally hidden behind stone or brick reveals, leaving little margin of the frame to be seen externally, this was to protect vulnerable wood from fire, which periodically ravaged parts of the rapidly growing cities. Glass, clearly a key component of any window, was predominantly "crown glass" the name derived from the manufacturing process which involved spinning out a globe of blown glass to form a disc which was then cooled on a bed of sand. Panes cut from this could generally not economically be more that 10 x 15 inches. For most of the century America relied on crown glass imported from Britain and it was not until 1787 that the first American window glass factory was set up in Boston. By 1830 glass-making had become very much more refined and 1832 saw the first plate glass produced. This revolutionized glass-making with dramatically larger panes, which in turn had a profound impact on window making. The substantial glazing bars with "ovolo" moldings needed to support small panes of vulnerable glass were no longer needed and were gradually replaced by slender moldings such as the popular "lambs tongue" Windows of the Regency period had significantly fewer glazing bars and by the 1840s sash windows could be made with no glazing bars at all. As areas of window glass grew so too did the size of windows, which often extended down to the ground to give access to the garden or at first floor level, verandas.

The Regency period also saw narrow panes introduced to the borders of window glazing, often with colored glass; the architect Sir John Soane delighted in using amber-colored glass at every opportunity.

Whilst windows give balance to the façade of a building, doors are the focal point and the 18th century saw the creation of the paneled door; prior to this most doors were of wide planks on heavy framing. Early Georgian doors were often six paneled with simple bold moldings, although by the Regency period the number of panels had reduced to two or three with subtle moldings and elaborate door furniture. External doors were invariably emphasized by a surrounding doorcase. Again designs became more refined as the century progressed: heavy hoods common early in the Georgian period gave way to simple pilasters supporting a pediment or lintel.

If externally millwork served to reinforce classical proportion then internally millwork gave owners the opportunity to show off their taste and sophistication and, of course, their wealth. Newly-discovered exotic hardwoods could be used to make doors, libraries, and staircases and the finest wood carvers could be employed to create ornately detailed fireplaces and doorcases. The most impressive millwork was unsurprisingly to be found in the principal reception rooms and in chosen rooms on the bedroom floors of a newly constructed early-Georgian house. The entrance hall and rooms running off the hall where guests were received and entertained were often paneled to their full height up to a cornice or crown mould. Millwork was generally robustly detailed with deep three-dimensional moldings such as the popular "egg and dart." Much of the paneling was in painted pine with occasional rooms, in Britain, in locally sourced oak. Doors to rooms on the principal floors in larger houses were invariably six paneled and of hardwood, wax polished to show off beautiful timbers and fine craftsmanship to best effect. The refinement of the Classical style with a movement towards finer, lighter details as the century progressed was seen in

One of SYMM's classically trained craftsmen applying the finishing wax to a Doric entablature crafted from English Oak.

the architectural detailing of joinery inside the house. By the 1760s the practice of full-height wood paneling was giving way to simply detailed wainscot, referred to as dado in England, paneling at low level with flat plaster on the walls above and, finally, the restraint of the Regency period saw plastered walls with only simply molded shallow profiled wainscot rails.

Whilst the role of joinery in ornamenting the Georgian house gradually diminished through the 18th century, staircases remained the most impressive feature of any well-to-do house throughout the period; here the millworker had a real chance to shine. Early Georgian staircases were robustly detailed with handsome turned balusters and heavy handrails, often constructed entirely from oak or even West Indian mahogany. Lesser homes used cheaper pine, but invariably still had hardwood handrails. The most impressive early Georgian staircases had open strings (where the ends of the treads are left showing), which were beautifully carved. Each tread would then typically have three balusters emulating columns sitting on urns separated by squared blocks. As the century progressed, balusters became more slender and sometimes twisted. These magnificent structures, befitting their importance, were well lit by windows and, later, skylights. In the last decades of the 18th century turned balusters were distinctly out of fashion and were replaced by thin "stick" balusters and by 1800 wrought iron increasingly took their place; the "age of wood" was coming to a close.

The Georgian period was a remarkable one, with building work on a scale not seen before, and those working in stone, brick, plaster, and especially wood dramatically developed and refined their craft.[1]

Remarkably, the craft skills that created these buildings largely survive today in both Britain and America. The growth of the conservation movement over the last 30 years, coupled with a revived interest in constructing new buildings, particularly those in the Classical style, using traditional craft methods and correct historical detail have created a strong demand for craftsmanship.

With enlightened patrons commissioning extraordinary work and architects and designers producing both scholarly and innovative design incorporating high craft, long may these skills flourish!

---

1 For further detailed information on the Georgian period I thoroughly recommend *The Georgian House in Britain and America* by Steven Parissien, from which many of the key facts in this essay have been drawn.

OPPOSITE:
Designed by Robert Franklin, the library at Hanwell Castle is fabricated in European Oak at SYMM's workshops in Oxford, and includes corbels that are hand-carved with fruit and flowers.

BELOW:
A hand-carved decorative wood panel
SYMM

A cast bronze finial, painted black, caps the newel post of a decorative French-inspired hand-forged stair.
Gold Coast Metal Works

Metalwork: The Jewelry on a House

Alain Olivier

金属制品：房屋佩戴的首饰

CLASSICAL ARCHITECTURAL METALWORK is experiencing a resurgence due to interest and demand from discerning homeowners and architects who wish to enhance the beauty of their homes and serve as modern-day patrons of high craft and the decorative arts.

A way to peer into someone's mind is to look at their bookshelf. Having spent the greater part of the past decade visiting architects' offices in the US and Europe, I am frequently gratified to find books on classical ironworking in their design libraries, such as surveys from F. Contet and Antoine Durenne or Louis Perroux's *La Serrurerie d'Art*. These volumes present stunning architectural ironwork from the 17th through 19th centuries—elaborate gates, grills, doors, and railings adorning palaces, cathedrals, universities, and private estates.

Many architects are well versed in the visual history of ironwork and cite outstanding artists like Jean Tijou and Samuel Yellin. However, many of these same architects are unaware of a new generation of highly-skilled artisan blacksmiths and metalworkers dedicated to preserving the craft today. Since architectural study no longer requires internships at craft studios and most blacksmithing workshops are located outside of urban centers, few architects have the opportunity to see these contemporary tradesmen at work and appreciate their ability to render industrial metals into lyrical forms and structures. The core of high-end classical architectural ironwork relies on blacksmiths who have completed a decade-long training process that takes them from apprentice to journeyman to master in a progression not much different than medieval European Guilds.

Metalworking ateliers are multidisciplinary operations, *manufactures* that produce their designs from conception to final installation. Fine ironwork unites the skills of engineers, draftsmen, moldmakers, blacksmiths, *repoussé* artists, *chasers*, and locksmiths. Blacksmiths still make their own tools and forge hot iron with their hammers at the anvil. The heat produces slag—an outer coat derived from the metal that defines its surface finish—and every hammer blow leaves impressions from the chisel while bringing the piece one step closer to its desired shape. *Repoussé* experts in the French tradition use their hammers to "push" out thin sheets of metal to form ornamental leaves, while those schooled in the Central European leaf-making tradition fashion leaves hot on the forge using a slightly thicker plate. When performed by a master craftsman, both methods result in the lifelike ornamental acanthus leaves that make traditional grills and railings so compelling to the eye. *Chasers* use their finely made mallets and chisels to incise bronze ornaments and rosettes by carefully engraving the surface. Each mark is a testament to the hand artistry, adding to the resonance of the finished piece. Locksmiths assemble the often thousands of leaves and scrolls in railings that can stretch for 50 meters, using everything from the classical assembly techniques of pins, rivets, and collars to the modern methods of soldering, brazing and welding. Moldmakers hand-carve shapes out of clay, wood or wax that are used to make models for cast iron or bronze ornaments.

Of course, time has not stood still and technological innovations have improved the preliminary tasks of measurement and modeling. Drafting departments take hand-drawn sketches or even pictures of the "sketch-in-iron" and through the wizardry of Photoshop make lifelike 3D renderings and photorealistic imaging.

While many layman still use the term "wrought iron" to refer to our trade, real wrought iron—a soft, easily worked, low-carbon iron with great plasticity—has not been produced for 50 years. It is still sought-after for restoration work, but can only be obtained on a reconstituted basis from a few companies in the European Union. Today, mild steel predominates and mimics the look and feel of its wrought iron antecedent. Other materials have also been added to our craft's toolkit. Stainless steel is forgeable and its unparalleled corrosion resistance and clean surface finish make it ideal for waterfront applications. Non-ferrous materials, such as bronze, can be cast forged and machined in a classic blacksmithing atelier. With special dies, these metals can be made into extrusions allowing for a whole new range of possibilities, from complete doors to applied profiles. For those wanting something even more novel, some in our trade work with titanium and innovative composite materials.

The past two decades represent a renaissance for classical metal work. The renewed interest in classical architecture, coupled with aesthetes willing and able to undertake the considerable time and expense to create a truly great house, is an opportunity for all in our trade. Political changes also allow a vast swath of Europe to re-emerge on the world market and re-introduce their centuries-old craft traditions to new discerning patrons.

Few craft elements enhance their settings as much as metalwork. Whether it's a chateau in Europe or a cottage in Newport, guests are often first greeted by a wrought-iron gate followed by a bronze door. Once inside, a hand-forged elegant stair railing sweeps through to define the airy space. Understanding the power of grand gestures, classical architects and their patrons gave these architectural flourishes the ultimate pride of placement—at the front of their finest commissions.

In an age of prefabrication and shortcuts, there still exists a small group of ateliers dedicated to maintaining the decorative art tradition. These highly skilled craftsmen, draftsmen, and engineers recreate the classical ironwork of past centuries to imbue a rich character to the contemporary Classic house.

Architectural flourishes on a hand-forged stair railing
Gold Coast Metal Works

The delicate decorative leaves on the stair railing of this Palm Beach residence, designed by Smith Architectural Group, are hand-forged in mild steel.
Gold Coast Metal Works

This Georgian chimneypiece displays the popular 18th-century device of inlaying contrasting marble within a field of white statuary marble.
Chesney's

The Centerpiece of a Room

Paul Chesney

房间的中心装饰品

IN WRITING ABOUT FIREPLACES it is necessary to first define the word as, over the centuries, it has assumed a wide variety of meanings in terms of both form and function. In the earliest times, it would have signified the fire that burnt on the earthen floor in the middle of a communal hall, the wood smoke generated by it making its way out through a hole in the roof. By Norman times, it was a simple arched brick and lime chamber built into the depth of a wall with only the most rudimentary chimney arrangement. The Tudor and Jacobean periods saw the design of fireplaces assume much greater importance with the fire surround as a decorative feature, often with an elaborate over-mantle carved in timber or stone incorporating heraldic detail and dominating the interior of a room. The importance of the fireplace as the focal point of the room continued during the Caroline period and later 17th century as Inigo Jones and Christopher Wren applied their knowledge and understanding of Italian architecture to fireplace design. The 18th century saw Baroque, Palladian, Rococo, and neo-Classical design all enjoy periods of popularity which were reflected in fireplace design and during this time an increasingly sophisticated approach was taken towards improving the functionality of the fireplace as an efficient means of providing heat.

Throughout these periods, the one constant element in fireplace design has been for a receptacle in which or on which to burn a fire. Thus the term fireplace can be taken to mean a set of fire dogs or andirons of literally any period, as the earliest examples of these date back as far as 300 AD in the case of a double andiron unearthed in Colchester, Essex. Alternatively it can mean a free-standing grate, a register grate, a stove or a finely carved timber, marble or stone surround—and this is by no means an exhaustive list. But for the purposes of this essay I am going to restrict the term *fireplace* to the designs of a single period, making reference only to the later Georgian period and to limit it further by reference only to the decorative surround that frames the fire chamber. I have selected this period as I believe that it represents the apotheosis of fireplace design in terms of variety, craftsmanship, and creativity. It is no coincidence that today this is the most widely copied period of Classical fireplace design and that examples of neo-Classical fireplaces are regularly commissioned for both historic and contemporary interiors.

Robert Adam is rightly credited with the popularisation of the style that is known as neo-Classical, and which defined the later Georgian period. Returning to England from Italy in 1758, he introduced a style of architecture and interior decoration based on Roman classical architecture but which interpreted and applied this classicism in a light and playful manner. Sir John Soane described Adam's work as a "revolution in art" and unquestionably it influenced every aspect of interior decoration. Adam devoted much time to applying this new form to the design of fireplaces, recognizing, like all great architects, that a successfully executed fireplace is key to the balance and symmetry of a room. There were features of Adam's work that separated it from the severe classicism of his predecessors and which allowed him to create a unique range of fireplace designs. Almost always relying on the highest standards of craftsmanship to achieve his effects, Adam introduced decorative elements to his work that were entirely new and original. Perhaps as a consequence of the enduring popularity of neo-Classical fireplace design, it is still possible today to find craftsmen who are able to execute this type of work. One of the most recognizable features of many neo-Classical fireplaces is the appearance of delicate low-relief carved ornamentation based on original Roman decoration. In today's age of mass production, encouragingly there still remain highly skilled marble sculptors and carvers who can successfully execute the low-relief trails and scrolls and bas-reliefs of classical figures required for a neo-Classical chimneypiece. There are no short-cuts in this type of work and the techniques and tools have changed little over the past 200 years.

Vibrant use of color is another characteristic feature of fireplace design that emerged during the period. Adam's approach was to use statuary marble not only as a medium for fine carved detail but also as a base for the inlay of a variety of richly colored marbles and semi-precious stone. Purists of the period were offended by this development but it added a new dimension to fireplace design with the rich color offered by intricately inlaid patterns of marble such as convent sienna, Sicilian jasper and Spanish brocatelle enhancing the visual impact and presence of many fireplaces of the period. The art of the inlayer is still with us and it is possible to replicate the most complex examples of inlay, providing the original material is still being quarried. A number of these polychrome fireplaces were also executed in a medium called scagliola. This was an Italian invention of a composite substance comprising marble dust and gypsum, which when held together by an adhesive, could be polished to imitate marble. More flexible and delicate than marble inlay, scagliola was used in many fireplaces of the period and continues to be accessible as a decorative medium through specialist workshops.

The broad church of ornamental devices that Adam employed influenced others during the period. Wedgwood panels became a popular decorative device within fireplaces and one Mrs. Eleanor Coade invented her faux stone material, Coadestone, which was used to manufacture a wide variety of classical ornament available for incorporation in fireplaces.

There are many other examples of new and original fireplace designs that emerged during this period, a number of which are still replicated today, but there is insufficient space to list these. What is unarguable, however, is that the best of neo-Classical fireplace design finds a place in many interior schemes today increasingly in the form of commissioned facsimiles rather than expensive antique pieces, the former relying on the same techniques and skills that produced the latter.

OPPOSITE:
The design of this Palladian chimneypiece includes contrasting inlaid marbles, obelisk jambs, and a carved centre tablet.
Chesney's

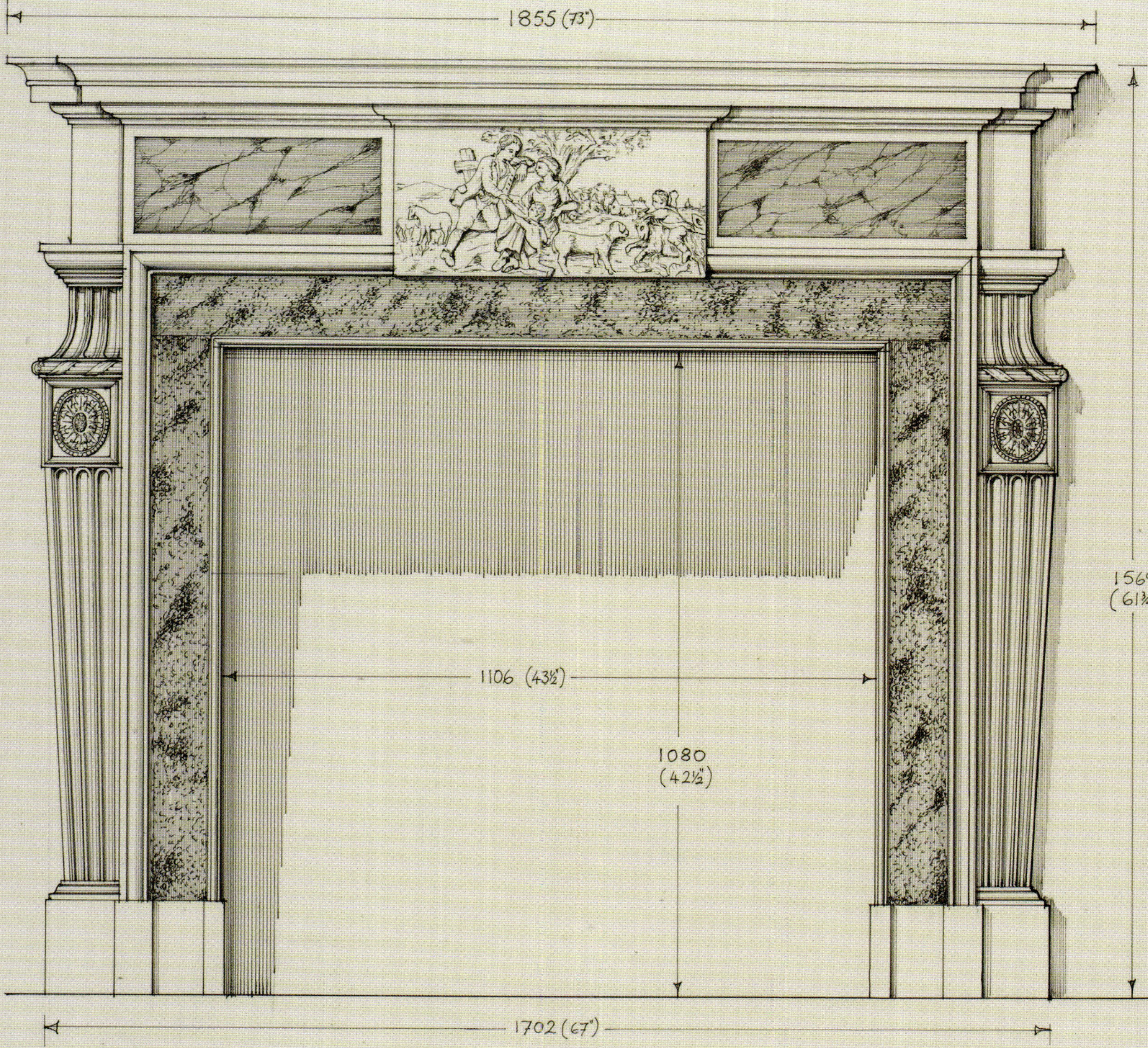
1855 (73")
1569 (61¾")
1106 (43½")
1080 (42½")
1702 (67")

A detail taken from a series of specially commissioned gilded lacquer panels, set within architectural wall panels
Kevin Cross Studio

Decorative Painting and Gilding: Classical Origins and Exotic Influences

Kevin Cross

装饰画和镀金画框：古典起源与外来影响

WHEN WE THINK of Classical art and architecture in terms of color, our experience is likely to evoke images in whites and grays, no doubt due to the survival of stone artifacts, and the loss of other objects and surface treatments created with less durable materials. Though there is evidence to support the assertion that some sculpture and even some buildings were originally painted and gilded, a preference for the notion of a monochromatic architectural environment in antiquity prevails to some extent, at least for exteriors.

If we put that discussion aside and shift our focus to exploring the question of artistic authenticity in interiors, we find that we have much more information to consider. Near the end of the 15th century, the ruins of the *Domus Aurea*, Nero's 1st-century Roman villa, were discovered. Upon excavation, the preserved decoration in stucco and fresco paint had a revelatory impact on the artists of the day. The work of Raphael and his team at the Vatican demonstrates their rapid assimilation of the vocabulary and movement of classical ornament; the Loggetta and Loggia are often cited as seminal examples of the classical grotesque style. As perhaps even greater evidence of their mastery, in the refined elegance of the Villa Madama we have not only an authentic interpretation of the past; we also have a sense of the future direction this style would take.

In the mid-18th century, the excavations of Herculaneum and Pompeii exposed a wealth of historical material, thus deepening our insight into everyday life in antiquity. From direct observation of these and other sites, many artists and architects who were on the Grand Tour at this time returned home with fresh ideas about adapting classical design elements to their own endeavors. William Hamilton's research and collections stimulated antiquarian interest in the area and contributed to what was called the Etruscan style. James Stuart's publications and works distinguish him as one of the early pioneers of neo-Classicism, but it is probably Robert Adam to whom we attribute the highest achievements in synthesizing all of the elements of classical architecture and interior design into a cohesive whole. In his buildings and interiors, in his publications, and most notably in his drawings, we see his attention to every detail, from the placement of the colors and gilding on the ceilings and walls to the design of the furniture that would ultimately occupy the space.

Alongside the continuing development of the neo-Classical style there were other influences on architecture and interiors that made their appearance, though none of these would ever achieve a dominant position, especially in reference to domestic architecture. We have Gothic Revival efforts such as Horace Walpole's Strawberry Hill, William Beckford's Fonthill Abbey and the works of A. W. N. Pugin. There was also an interest in the art and architecture of Egypt, fostered by Napoleon's campaign and the subsequent scholarly publications. From these sources came the inspiration for some aspects of the Empire style of Percier and Fontaine, and the works of Thomas Hope.

One of the exotic influences that might claim the most lasting effect is that of the Far East. Over the Silk Road, art objects and designs from Asia had been entering Europe intermittently for centuries, primarily through Venice. That changed forever when, in 1497–99, Vasco da Gama sailed around Africa to India and back, proving that a sea route to the East was viable. By 1543, Portuguese mariners had reached Japan. Without delay, other European nations also turned their naval attention eastward and by the early 1600s we see the establishment of their individual East India Companies. Though the initial goal of exploration was primarily to discover the source of spices, interest in the other cargo of the returning ships, i.e. porcelain, silk, tea, and lacquer, proved so great that intense effort was soon under way to expand the market.

By the middle of the 17th century, the emerging impact of Asian art and design, particularly that of China and Japan, was felt all over Europe. The effect of this generated an enthusiasm for all sorts of Oriental exotica: architectural follies were constructed, gardens were designed in the informal Chinese style, and many stately houses had rooms configured in a fanciful Oriental fashion. Initially, export art such as Chinese screens, both in the colorful Coromandel-style lacquer and in gilded black or red lacquer, were dismantled and installed on walls to create a "lacquer cabinet," which served to display the owner's connoisseurship. With the demand exceeding the supply, skilled artisans soon began to imitate the decorative effect of lacquer by using traditional painting and gilding materials, to either accurately supplement the original panels to complete a room, or to create a room entirely in a European version of lacquer. To facilitate this, treatises were published with detailed instructions on how to create a lacquer finish to rival that of the Eastern ware, while many other books with engravings of ornamental designs provided suggestions for composition. Eventually the word "Chinoiserie" was created to describe this phenomenon; this term is now commonly used to encompass almost all aspects of Far Eastern artistic influence, from its earliest manifestations to those of the present day.

Although, for clarity, we tend to examine these various trends as separate, there are many interesting instances of coexistence. To cite just a few, we have the marvelous Rococo arabesques of Watteau and Huet, William Chambers' expertise in neo-Classical architecture and Chinese design, and the outstanding collection of Japanese lacquer amassed by Beckford and displayed at Fonthill. Even in more recent times, the taste for combining distinct artistic styles is apparent, as we see in the placement of Japanese art within the interiors of Frank Lloyd Wright, and the presentation of antiquities in modern settings.

Given the popularity of period rooms in major art museums and the consistent attendance of house museums, from Caramoor to Hearst Castle, it is evident that many people enjoy seeing a wide variety of interiors. A glance at the pages of any book or magazine on the subject of architecture and interiors will reveal, side by side with examples of contemporary design, many illustrations of historical decorative treatments, such as landscape murals, Chinese wallpaper, classical grotesques and arabesques, Venetian plaster, gilded composition ornaments, lacquer panels, *trompe l'oeil*, faux finishes of wood and marble, and so on. From this we can conclude that the ancient desire to have color and texture in our homes is still alive, and the requisite capability of designers and artisans to satisfy that demand is flourishing.

A pair of painted Philadelphia chairs and a Baltimore table, painted black with gilt polychrome decoration, ornamented by Kevin Cross

RAVENWOOD (see page 141): Modeled by Foster Reeve & Associates, and carved in Mara Crème limestone by Traditional Cut Stone, the entrance surround includes a richly embellished entablature that is topped with a 2-foot-tall raven.
*Designed by Richard Cameron**

The Permanence of Stone

Richard Carbino

石之永恒

STONE IS THE MOST WIDELY USED building material known to man. Its workability, durability, abundant supply, and inherent natural beauty have made it the top choice for buildings, monuments, sculpture and architectural ornament since the beginning of civilization.

Man fears nothing but Time; but Time... fears only the Pyramids. This ancient proverb is perfect, even before the multitude of uses for stone are expressed, one thing though is noted, the material's inherent *presence.* The pyramids—mountains of endless stone, stood tall and outlasted the Empire of the Pharaohs they entombed, striking awe and intimidation into those who saw them.

Simply stated, stone has always been here and always will be here. Its life cycle is that of the earth itself, and Man has paid homage to that, time and time again, since the moment we started using it to build. And stone is indeed one of history's most commonly used materials; it started with the pyramids and continued with the Great Wall of China, the cathedrals of Europe, the soaring skyscrapers of Manhattan and the mansions and humble abodes that cover the earth. Though initially the deafening silence of stone intimidated many, it soon became the fascination and obsession of many artists and builders who wished to harness its beauty; who dared to work *with* it, to break it, and shape it to create and document details of the imagination that would last for many lifetimes to come. It truly is the material that the book of human history has been transcribed upon.

Today, I pay homage to stone by fusing state of the art technology with the expertise of classically trained stone carvers in an attempt to keep the ancient art alive and well in this crazy, fast paced modern world in which we live. New architectural ornamentation, figurative statuary and heritage restoration are my three favorite subjects. Limestone, marble and sandstone are my three favorite mediums. I believe, people are still moved when they realize they too have an opportunity to participate in personalizing something that will quite literally withstand the test of time and, in essence, render them immortal to some degree. In offering this service, I have an opportunity to pay respect to the craft, to art, to history, and to the material—all the while contributing to architecture, which will influence the way people live and love. Stone is used for sculptural work outside of architecture, of course, but its use as both a structural and decorative element in architecture is what has really pushed me to continue carrying the artistry of stone carving into the modern world.

Evolving with history, as all things do, the uses and refinement of stone have gradually been worked into architecture's own evolution. Yet, despite great technology around us, the process and even most of the tools used to carve and manipulate stone are the same today as they were centuries ago. The basic implements necessary consists of: a chisel, mallet, and an artist's sixth sense for communicating a vision onto stone.

In a broad sense, stone fabrication can be broken down into the following categories: building stone (rough stone assembled by stone masonry installers for building walls and walkways) and dimensional stone (stone that is cut to specific sizes by hand or machine for architectural elements or ornament). Two separate and distinct professional trades fall under the dimensional stone heading; they are that of the stonecutter and the stonecarver.

The stonecutter hand works stone into geometrical shapes and sizes from information (shop drawings and shop tickets) that are provided to him/her by the drafting department. Some items produced by a stonecutter would be: window sills, wall coping, window surrounds, cornices, and fireplaces. All of the information the stonecutter receives is in written form. This trade is more mathematical than artistic. The stonecarver can do everything a stonecutter does, but also has the ability to carve artistic elements in three dimensions. This can either be done free-form for sculptural or stand-alone elements, or will be done by replicating a three-dimensional plaster model that has been provided by the art studio. Some items produced by a stonecarver would be: gargoyles, Corinthian capitals, and figurative statuary.

The rise of popularity in glass curtain wall systems in the 1970s dealt a near fatal blow to the dimensional stone trade in North America (in the 1950s there were over 50,000 stonecutters and carvers in North America, today there are under 1,000). Nevertheless, despite all of this, the art of stone carving remains alive and well and is, thankfully, growing once again in popularity as people are demanding that natural materials be used on their projects.

Architectural ornamentation speaks simultaneously to the style of architecture and to the cultural state of society—at the time of construction. After a generation of looking upon everything as a disposable commodity, people are now weary of continually throwing out the old and purchasing the new. More and more buildings are once again being designed with a sense of permanence—as testified by the projects shown in this book. And stone—as both a structural element and a blank canvas, is a must for anyone who desires to tell a story through architecture. One needs only to find a reputable stone carving shop in order to be able to have their dreams, inspirations, and personal history, given eternal life. It is human nature to want to stand up, be noticed, and not be forgotten.

The gargoyles that sit atop Notre Dame Cathedral in Paris, have quietly watched the world change for centuries and they will continue to watch over us, forever.

Rosettes, carved from limestone by *Traditional Cut Stone*, adorn the frieze of a bespoke fireplace.
Wadia Associates

A design rendering for a classical urn topped with a overflowing floral arrangement—later to be carved from limestone
Rendering by Lawrence Voaides, Traditional Cut Stone

A detail from a decorative plaster ceiling
Foster Reeve & Associates

Ornamental Plasterwork

Foster Reeve

装饰性石膏制品

ARCHITECTURAL ORNAMENT is my passion. Working with architects and designers to create beautiful works of art that adorn buildings is a challenge that is as rewarding as it is demanding. The process of developing the correct ornamentation for a given house or room is very subtle and complex. The team will often consist of the architect, designer, and homeowner, with one taking the principal role of communicating with the artist. When the team initiates the concept for an ornate moment, whether a simple molding or an entire ceiling, it is very helpful if they supply good reference imagery in order to set the tone for the work. These precedent images can provide guidance for both the detailing of the ornamentation, and for the feeling of the space as it is impacted by the decoration. Next, I like to draw from the source material, even slavishly copying details, in order to take ownership the imagery and understand the parts or aspects that apply to the current work. This is a great process of discovery which often surprises artists with information that they didn't know was there before, as well as confirming the hierarchy of what is truly important about the reference material. In this way, copying from great work is probably one of the most valuable activities in the process of developing original art.

Once I feel comfortable with the imagery, I like to make a lot of quick sketches of ideas for the work at hand. At this point in the process, a meeting with the design team usually produces enough feedback to proceed with developing a couple of detailed drawings which are again shared with the team, and we start to hone in on what will be the final design. Once the design has been settled on, final measurements are established, modules of repetitive ornament are fixed, and a commitment is made to begin sculpting.

The reference material again comes back into play with the sculpture. The line, texture, massing, and overall feeling of the work needs to be clearly understood. Sketches in clay at full scale, along with the precedents, provide the team with a clear vision of the work. The work in progress is reviewed by the team, and often digital images are sent when studio visits are not possible. Indeed, we have produced entire complex projects with all reviews being done by photograph. Once the final artwork is finished, it is time for the molds to be made and products to be cast and installed.

Gypsum plaster is unparalleled in its use for interior architectural ornament. It can be drawn, carved, and cast into a nearly unlimited variety of forms. Plaster's strength and malleability have made it the best material for achieving the complex forms and details of classical decoration.

The use of gypsum plaster predates recorded history. The Egyptians were very familiar with gypsum, and used it to coat and decorate the interior walls of many of the pyramids. It has been used ever since to produce beautiful product for interior architecture.

Gypsum is a naturally occurring mineral compound that can be crushed and heated to produce what we commonly call "plaster of Paris." The term was coined in the 18th century for the high-quality product that was, and still is, produced in the Paris region which set the gold standard for purity and strength. When mixed with water, gypsum can be molded or applied to form almost any shape or texture. Historically and indeed continuing to day, this versatility provided many opportunities for innovation in design.

The wealth and culture of the Renaissance provided fertile ground and for the increased use of gypsum plaster in its many forms. Beginning with the Counter-Reformation, when church building took on a particular sense of urgency, gypsum was employed in the creation of vaulted ceilings supported by timber framed roof assemblies. The stone vaults of the Middle Ages were thus replaced with faster and cheaper materials, while still providing the grandeur associated with a temple.

As the craft of plasterwork developed, the burgeoning upper classes soon took advantage of the opportunity for opulence in their private homes. Classical Greek and Roman architecture provided a treasure trove of design inspiration. Gypsum was a perfect choice for creating shape and ornament of unrivaled complexity and beauty to adorn the neoclassical edifice.

I must differentiate here between gypsum plaster and lime plaster, as both were used to create the "plaster" interiors of the Renaissance. Lime, the basis of lime plaster, is the byproduct of heating limestone in a kiln to drive off the free oxygen. The cooked limestone is then crushed into a powder, which, when mixed with water, produces a putty that can be combined with graded sand to produce mortars and stucco renders. Lime putty does not set hard on its own, but is a relatively stiff and sticky "stuff" that can be combined with gypsum to form moldings by dragging a profile knife along a track. A classic cornice molding of the time would have been produced in place using this technique, with the addition of cast gypsum ornament laid on by hand.

Today, rubber mold technology enables casting entire moldings in gypsum. The installation of these casting goes much faster and with less jobsite mess, and does not require the special skills of the run-in-place technique.

Cast and formed gypsum provides designers the opportunity to explore architectural decoration with unparalleled versatility. From highly abstract geometry to mimic cultural symbolism and storytelling, gypsum is the material that can ornament and add meaning to the human space.

A scagliola panel with plaster ornamentation.
Foster Reeve & Associates

A craftsman hand-modeling a Corinthian capital in clay that will later be used to create a mold to cast the finished piece in plaster.
Foster Reeve & Associates

During the restoration of Hanwell Castle, H. G. Matthews Brickworks was tasked with fabricating new handmade bricks to match originals. Lime mortar was then added to blend in with the existing pointing.
Robert Franklin Architect and SYMM

The Beauty of Handmade Bricks

Jim Matthews

手工砖之美

BEFORE BRICKMAKING machinery began to be developed in the 19th century, all bricks were made by hand. A skilled brickmaker would be expected to make at least 1,000 bricks a day but there was far more work involved than simply throwing the clay into the mold.

Before he could begin making bricks, the brickmaker would first have to prepare the clay, making it into a workable consistency by mixing it with water, this was known as pugging. The earliest method of doing this was to tread the clay with bare feet, later pug mills were developed which had a vertical shaft with blades to mix the clay and the shaft was turned by a horse, which was harnessed to an attached horizontal pole. The pug mill had to be loaded by tipping wheelbarrows full of raw clay into it. The mill would usually be sited above the molder's bench so that the pugged clay would fall down to where it was needed and the arduous task of the brick molding could begin. Once molded, the wet bricks were placed on a long wheelbarrow and then pushed to the drying area, where, by holding the wet brick between two wooden boards, he would carefully set them on edge onto the ground to dry.

Freshly-made bricks each contain around a pint of water, which has to be dried out before firing. They were therefore placed outside in rows, called hacks, one course high at first and then slowly stacked up as they dried and could support more weight. They were also turned every few days at an angle to allow the wind to pass through them more easily. When dry they would be wheelbarrowed to the kiln to be set in place for firing. Firing would take place day and night over a number of days using wood, and in later times coal, as the fuel. The bricks would then be taken from the kiln and wheelbarrowed to the stacking area where they would be kept until sold, and hand-loaded onto a cart to be horse driven to site. The bricks would then be laid in lime mortar, the production of which, at the time, was another heavy manual process.

Considering before any of this happened the clay had to be first dug by hand, it is worth remembering when looking at an old brick building, the staggering amount of physical effort that went into producing it.

Eventually mechanization meant enormous labor savings were possible, and machine-made bricks could be produced much more cheaply than handmade bricks. The early machine-made bricks were also esteemed for their regularity and neatness and were considered advanced and refined, which was a novelty when compared to the rustic handmade. Slowly the small handmade yards were driven out of business and replaced by far fewer but far larger mechanized yards. This process was accelerated by the parallel development of road and rail networks, which made the transport of bricks relatively easy and cheap, even over large distances.

The small yards that relied on outside drying could not risk putting bricks out from November to April due to the risk of frost ruining the wet bricks. In the winter when brickmaking was not possible clay was dug for the following season. The larger yards with their greater resources were able to invest in indoor drying systems that made brickmaking possible 12 months a year putting further pressure on the small yards.

This trend towards large-scale production at the expense of small local production continued relentlessly in England, where in 1850 there were around 2,000 brickworks but there are now less than 50, with around 10 producing handmade bricks. This has had the desperately sad effect of blurring the local distinctiveness of buildings that used to be such a delightful feature of the landscape. Once it was as if buildings were, in effect, the ground turned inside out, with houses reflecting the geology beneath them with different clays producing different colored bricks. This is now rarely the case but perhaps in the future, environmental concerns and a renewed interest in localism will tilt building culture back in favor of local materials.

In spite of the cost advantage of mass-produced machine-made bricks, handmade bricks have never been completely replaced because there have always been people prepared to pay a premium for their beauty and character. The essence of a handmade brick is its texture, or the sand crease. This is either absent in machine-made bricks and artificial attempts to reproduce the sand crease mechanically are never quite the same. The sand crease is formed by the clot of clay being rolled in sand by the brick maker before it is thrown into the mold. The primary reason for this is to stop the clay sticking to the mold, but a favorable side effect is that the excess sand falls away when the brick dries, leaving an impression behind. Each maker will have a slightly different technique using more or less sand, creating finer or deeper creases. This means the bricks have a character unique to this process, which is simply absent when bricks are made mechanically. This subtle but important difference has enabled a small number of traditional yards to survive.

The traditional yards also had the advantage that old buildings made out of handmade bricks would always need materials produced in the same way if they were to be repaired and conserved correctly. Increasingly strict control has ensured that this is now nearly always done, and had provided a niche, which has helped keep traditional brickmaking alive. The manufacture of small runs of bespoke specials is something that particularly suits small yards; a large, volume-focused yard, would find such work too disruptive to their production.

The production of gauged and rubbed work is a further example high-quality production that takes place on a specialized small scale. It usually involves the use of refined clays which have had all trace of impurities removed by filtration or settlement processes. This clay is molded into a larger than a standard brick size, which, when fired, can be rubbed to form very precise bricks with sharp arises or cut with wire bow into elaborate shapes. They are also used for carvings and sculptures.

So, despite the revolution in technology and distribution that has changed the economics of production, the beauty of handmade bricks means they will never become obsolete.

OPPOSITE:
A stack of handmade bricks show how each one varies slightly from the next, creating the richness that is sought after with this building material.
H. G. Matthews Brickworks

"We shape our dwellings, and afterwards our dwellings shape us."
Winston Churchill

项目集
THE PROJECTS

A New Country House

HAMPSHIRE, ENGLAND

Robert Adam, ADAM Architecture 新乡村风格别墅

THIS NEW COUNTRY HOUSE, within a 700-acre working farm in Hampshire, is a highly individual classical design that features fine detailing throughout. The project was the first to be won on appeal, and built on a *greenfield* site, under the British Government's policy of only allowing the construction of new country houses that are proven to be of the highest quality.

Robert Adam designed the house with a consistent and clear geometry, including elevational treatments that reflect the different functions of the rooms within the house. The formal wing of the house accommodates the principal public rooms, set within a square plan. The elevations are ordered by the use of paired stylized Corinthian pilasters, with a balustraded parapet, and a secondary Doric order on the ground floor. This wing is linked to a square tower with a shallow copper dome that contains the farm office and master bedroom above. The tower uses the proportions of the giant and secondary orders as the formal wing but suppresses the pilasters, making it visually subservient to the formal wing. The formal section utilizes large windows and classical detailing influenced by the classical architecture of Alexander "Greek" Thomson and Karl Friedrich Schinkel.

The family wing is on a separate square plan with an open courtyard. It has reduced ceiling heights and its elevations are based on a giant order with paired shallow pilasters, reduced in scale and detail from the giant order on the formal wing.

The use of materials and detailing visually reinforces the hierarchy of the two wings of the house. The importance of the formal wing and tower is expressed by the use of stone, whereas the less formal family wing is faced in brick. The light cream color brick blends with the bath stone of the adjacent stone wing, but the differences in textures create subtle and attractive contrast.

DECORATION: AN ECLECTIC LEGACY
In his early career, eminent Scottish architect Alexander Thompson developed his own style of decoration adapted from Greek, Egyptian, and Levantine sources. Here we see the influence of that eclectic style, as architect Robert Adam has adorned this new country house with Greek-inspired anthemion, Egyptian palmiform capitals, and classical urns displayed on top of engaged Greek Doric columns flanking the entrance. Derived from the honeysuckle flower, the anthemion along with the acanthus leaves are the predominant natural forms of decoration in classical decoration. Used in antiquity to contain cremated remains, urns later became popular first during the Renaissance, and then during the neo-Classical movement as architectural and landscape decoration.

"Classical architecture has always responded to the changing needs of society. The opportunities offered by technological innovation and construction techniques are just some of the new demands that bring about changes in a living tradition."
Robert Adam

Henbury Hall

CHESHIRE, ENGLAND

Julian Bicknell & Associates 亨伯里庄园

HENBURY HALL is a new building in the midst of a beautifully landscaped estate dating from the 17th century. The original house had been dismantled in the 1950s but the original stables and cellars survived.

Julian Bicknell's design is based on an original concept by the painter Felix Kelly—Vanbrugh's Temple of the Four Winds at Castle Howard meets Palladio's Villa Rotunda. The principal rooms on the raised *piano nobile* are arranged around a vaulted central hall running from end to end of the house to the other, and rising 50 feet through the second floor, containing the bedrooms, into the dome. The Ionic order of the Serlian windows at either end of the hall links the interior to the exterior architecture.

The pediment sculptures were carved by Simon Verity, and the reconstituted stone carvings, including the Ionic capitals and decorative window surrounds were cast from original molds by master craftsman Dick Reid. The plaster friezes and cornices were made by local craftsmen combining original designs with traditional forms and features—some copied from Castle Howard, where Julian Bicknell had worked a year ealier on the Garden Hall and Library in preparation for the TV series *Brideshead Revisited*. The interiors were developed and elegantly decorated by David Mlinaric using furniture and paintings already in the owner's collection against a background of traditional colors, fabrics, and wallpapers.

Besides the Great Hall, there are four reception rooms each with an elegant antique fireplace. The east-facing dining room with its baroque bracketed cornice is decorated with Venetian silk and features decorative wood doorcases hand-carved by Dick Reid and a blowsy Venetian chandelier. The west-facing drawing room with its full-blown Corinthian cornice is hung with fine paintings. The smaller morning room on the southeast corner and the study on the southwest corner have vaulted ceilings and are fitted out with cabinets and bookcases integrated with the architecture.

Further images from this project can be found on pages 26, 52, 68, 71 and 72.

PRECEDENT: LA ROTUNDA

The owners of Henbury Hall originally came from Venice, so it is no surprise that the design of their home should based upon Palladio's Villa Capra, better known as La Rotunda, in nearby Vicenza. Considered by many to be the quintessential idyllic country retreat, this is not the first time that La Rotunda has served as inspiration for an English Country House. Built in the 18th century, Mereworth Castle by Colen Campbell and Chiswick House by Lord Burlington, are two previous examples.

"The history of classical architecture is punctuated in every generation with examples of striking innovation in both form and detail."
Julian Bicknell

A Park Avenue Apartment

NEW YORK CITY, NEW YORK

Timothy Bryant Architect 公园大道公寓

TIMOTHY BRYANT'S restoration and renovation of 550 Park Avenue re-imagined the 1917 apartment as an elegant pre-war apartment. Wren-period Georgian profiles and James Gibbs-inspired cornices convey a loosely 1930s feel, made all the more authentic by the revised layout's traditional division of formal and private areas, with the foyer as the central circulation area—as original architect J. E. R. Carpenter intended.

In the entry hall, Doric entablature meets Mauch cornices, gold-leaf cove moldings and hand-printed Italian wallpaper. A floor of interlocking octagons, fabricated in water jet-cut stone, references Venice's Convento della Carita. To the south, the rebuilt library was handcrafted in Paris, in old-growth Oregon pine, and showcases a salvaged Georgian carved-pine mantle.

The client's remarkable art and furniture collections are front and center in every room, where ceramics, murals, rugs, and ancestral paintings are side by side with impeccably crafted new pieces. Against the depth and shadow of the dining room's highly embellished moldings and over doors, mid-20th-century Irish and Asian themes come together in the specially commissioned Chinese murals, Irish sideboard, ceramics, and Aubusson rug.

The same serendipitous mix is found in the living room, where Bryant's Corinthian entablature introduces the client's Georgian-style scagliola mantel and an antique mirror, purchased in Ireland more than 40 years ago.

From the refinished original Cuban mahogany floors, to the subtly distressed Georgian dining table, a love of details is evident throughout. A rhythm of wainscoting provides a sense of continuity between the main living spaces, while unique hardware pieces by Charles Edwards of England, such as the Regency-style ebonized beehive doorknobs, are eye-catching finishing touches.

A further image from this project can be found on page 38.

PRIORITIES: FOCUS OF ATTENTION
Self-editing is perhaps the hardest task for any designer. Some rooms allow you to use your full bag of tricks, whereas other rooms are best tackled with a light hand. This is perfectly highlighted in this Park Avenue apartment, where architect Timothy Bryant allows the architecture in the living room to take a back seat to the client's furniture collection—as well as the spectacular views of the City. However, in the dining room and library, where the view is not as prominent, the decorative palette is unleashed with Chinese murals, wood paneling, and intricate moldings embellishing the walls, and leaving no doubt to where the focus of attention is directed. Likewise, as the entrance hall is sparsely furnished the now un-obscured floor becomes the focal point, allowing the opportunity to design and incorporate a highly decorative polished marble floor.

"Classicism is the language of tradition. It is the foundation of everything."
Timothy Bryant

Ravenwood

NEWTOWN SQUARE, PENNSYLVANIA

Richard Cameron* 瑞文伍德庄园

THIS LARGE-SCALE PROJECT originated with a comprehensive master plan for the 125-acre property, designed in collaboration with landscape architect Barbara Paca. A new entrance sequence was devised with the creation of a new main gate and allée. The new drive gives a gradual introduction to the house through its fields and stands of mature trees. The original 15,000 square foot main house was completely renovated and a new courtyard, attached gatehouse (which includes a home office and guest quarters), a gallery, a double-height library, a motor court, and an underground theater were all added, bringing the area of the new house to an impressive 27,000 square feet.

Formal gardens were added to the west of the house and a secret garden was laid out around the existing pool house, which was renovated to become an art studio. A new pool pavilion and rose garden were built on the back lawn. All nine existing outbuildings, including several 18th- and 19th-century farmhouses, a large stable, and a former dairy barn, were all restored and renovated as detached guest quarters or service buildings.

While the owners loved the orientation and layout of the property, they were not happy with the 1930s neo-Georgian style of the existing house. As such, it was decided to design the additions in a way that made them appear older than the original building. Utilizing the concept of vernacular progression, the entire house would tell a story—where a series of older outbuildings had been unified when a large manor house was later built. For stylistic inspiration Richard Cameron studied "old additions" to buildings like Groombridge Place in Kent, the nearby Pennsbury, and in particular Folly Farm in Berkshire—where Lutyens added a vernacular Arts and Crafts wing to a more formal William and Mary style house, that he had designed some six years earlier.

Further images from this project can be found on pages 1, 88, 89 and 114.

**This project was designed by Richard Cameron while he was a principal at Atelier—The Art of Building, LLC.*

CRAFTSMEN COLLABORATION:
TRADITIONAL CUT STONE AND FOSTER REEVE & ASSOCIATES
Typical of English classical design, the front door is one of few locations to feature any ornament. The Mara Crème limestone door surround is custom carved by Traditional Cut Stone and, in keeping with the style of the house, is based upon English Baroque models. Sitting atop the entablature is a limestone raven, modeled first in plaster by Foster Reeve & Associates, and then carved by Traditional Cut Stone (see page 114).

"Architecture is an art first and foremost. For it to be classical it must—to quote Wren—'aim at eternity.' Anything else is building."
Richard Cameron

A Classical Pool Pavilion

HOUSTON, TEXAS

Curtis & Windham Architects 古典池阁

BUILT TO ACCOMPANY a historic residence designed by well-known Houston architect John Staub in 1936, this brick and limestone pavilion is the architectural focus in the backyard of this Houston, Texas property. The placement of the pavilion on the site is analogous to an orangery, located perpendicular to the existing house across a lawn and allée of live oak trees.

Clad in brick with a weathered paint finish, the material palette of the pavilion follows that of the house. Where accents are rendered in wood at the house however, the exterior ornament of carved limestone used on the pavilion hints at the building's more distinctive function. The pavilion's program consists of a large entertainment space along with generous service kitchen and storage areas, housed inside a structure that masks its size with carefully designed proportions and detail. By applying the flattened and attenuated ornament of the Regency style, the pavilion façade complements the facing elevation of the existing house. The layered ornament suggests the tectonics of the building, where brick arches are supported by carved limestone arches and keystones. Fluted detail at the exterior arches and a Greek key at the frieze hint at the geometric design of the interiors. Three arched openings dominate the exterior order, while on the inside, a plaster cornice anchoring a lofty cove and tray ceiling define the entertainment space.

While still remaining true to the proportions of the Regency style, the more restrained exterior gives way to a more exuberant interior: black-and-white marble inlay floors, and a plaster ceiling with floral embellishment, are evocative of the Art Deco period. Ultimately, the pavilion integrates two design concepts: understanding that all accessory buildings should remain architecturally subordinate to the main house, while also adhering to the program by creating a grand entertainment space that is resplendent in scale and ornament.

A further image from this project can be found on page 21.

ORNAMENT: REPETITIVE PATTERNS
There is a great variety of geometric decoration in classical design, some of which architects Curtis & Windham have used in embellishing this pool pavilion. On the exterior, the egg and dart motif from the echinus of the Ionic order is repeated on the pilasters; a Greek key fret pattern adorns the frieze; either side of the cushioned keystone are flutes that are carved into the arched window surround; and limestone plaques decorated with flutes and roundels are centered over each arch, as is a partial engaged balustrade at the parapet wall. Inside the pavilion the decoration around the window surround continues with reeds, roundels, and a single guilloche that terminates with an imbricated bellflower motif.

"We see ourselves as part of the continuum of architectural tradition and look to each project to build upon this rich history."
Bill Curtis & Russell Windham

Farmlands

COOPERSTOWN, NEW YORK

Fairfax & Sammons Architects 农田庄园

FAIRFAX & SAMMONS was commissioned to design a new house for the grandson of the founder of Anheuser-Busch, on the family hops farm in upstate New York. Built on the site of his grandparents' house, this new stone house was intended to become the new family home for generations to come.

Sited on the famous Glimmerglass Lake, it commands a breathtaking view across the water. The house is constructed of load-bearing local stone and is designed in a neo-Federal style, allowing the house to blend in with its neighbors. The main core of the house is constructed with the local New York bluestone, the flanking bays are wood framed with coursed shingles, and the roof is heavy slate. These materials are a feature of local buildings of the Federal Period. The use of local materials worked in a traditional manner gives the house its sense of durability and its feeling of belonging to the rugged local landscape.

Upon entering the 10,000-square-foot house, a two-story entrance gallery opens up to the second floor. The curved stair provides a graceful connection from each floor. Beyond the gallery lies a living room that is flanked by a pine-paneled library on the left and a formal dining room on the right. These three principal rooms, connected by an enfilade, have numerous large windows and French doors allowing for a panoramic view from the rear of the house, with the lake in the distance.

Inside, paneled rooms were carefully detailed with fine custom millwork throughout. The central portico contains a suspended balcony providing a porch for the master bedroom overlooking the lake. Palladian windows add a soft curve to a somewhat strong and dominant façade. In the wings, which open onto south-facing loggias, there is a guest bedroom suite on the east side and the kitchen and breakfast room on the west, with bedrooms on the floor above.

Further images from this project can be found on pages 22 and 238–239.

PERMANENCE:
USING LOCAL MATERIALS

The load-bearing masonry walls of Farmlands are constructed of 10–12-inch-thick local New York blue stone that was quarried near the site. The stone was pried from ledges and laid in lime-mortared coursed rubble, which gives the exterior its wonderful texture and variegated coloring. The quoins on the outside corners graduate and diminish in size from bottom to top, countering against perspective while reinforcing the perception of a grounded structural mass. This combination of local materials used in a traditional manner with subtle detailing gives the house its sense of belonging.

"Classicism, be it expressed informally or formally, in the domestic realm offers creative solutions, both timeless and beautiful."
Anne Fairfax & Richard Sammons

BREEDS OF CATTLE

Hollywood Regency

LOS ANGELES, CALIFORNIA

Ferguson & Shamamian Architects 好莱坞摄政风格别墅

THE PEDIGREE of this 1930s Hollywood Regency style home had more to do with its past owners—Fanny Brice, Alan Ladd, and Jerome Moss—than the architectural design of the house. Although the front façade of the home was charming enough, the interior construction was wafer thin, and resembled a movie set rather than a house. Choosing to retain the brick façade with its quoin details and swan-neck pedimented doorway, Ferguson & Shamamian were instructed to re-create the interiors of a great Los Angeles house of the period.

Everything beyond the front brick façade was demolished, and all but the foundation, floor plate, and curved staircase in the entrance hall were removed. With access to other significant houses in Los Angeles, details were documented, discussed, and adapted, so that the local interpretation of classicism could be respected. Formality is often casual in Los Angeles, and with this in mind the house plan is rather less rigid than normally found in a classically designed home.

On entering the house, the restored staircase curves gracefully, allowing the hosts to still make that grand Hollywood entrance. Demolishing the house allowed the luxury of raising the height of ceilings in the principal rooms, and lowering some elsewhere, therefore manipulating how the spaces are perceived, and creating a distinct character for each room. This is further highlighted by the different treatment of floor, wall, and ceiling surfaces—whether that be panels or mirrors on the walls, or beams, coffers, and vaults on the ceiling. The Great Room, which also serves as a home theatre, is the perfect example of this surface treatment as it features deep walnut paneled walls and a cove-vaulted ceiling, creating an old-world feel that is brooding and introspective.

A further image from this project can be found on page 40.

FIRST IMPRESSIONS:
CHOICE OF MATERIALS
The entrance hall is the first room that you experience when entering a home, and as such provides an opportunity to set the character for the entire house. As these rooms tend to be sparsely furnished, the floor often becomes the focal point. A highly-polished marble or terrazzo floor instantly creates an air of formality, whereas a reclaimed limestone or terra-cotta floor is far more rustic. Likewise, a wood parquet or herringbone pattern creates a dignified elegance, that is in contrast to the feeling of warmth and age that can be found in a simple hand-scraped wood plank. In this house Ferguson & Shamamian successfully capture the casual sophistication of the Hollywood Regency style by combining white painted wall panels with an intricate and eye-catching parquet floor made from oak, walnut, and ebony.

"We accept and carry forward the canons of architectural traditions. The past, however, is never copied but interpreted—it encourages continuity without discouraging innovation."
Mark Ferguson & Oscar Shamamian

Wayside Manor

THE PLAINS, VIRGINIA

Franck & Lohsen Architects 路边庄园

DESCRIBED BY COUNTRY LIFE MAGAZINE as "pure English Baroque" this new Manor house designed by Franck & Lohsen Architects is set within a 100-acre estate in the heart of the Virginia Hunt country outside Washington, DC.

Nestled within these rolling pastoral farmlands, the front entrance façade opens to a generous forecourt on axis with a grand allée—and in the distance a new architectural folly, also designed by Franck & Lohsen.The forecourt, with a lily pond at its center, is commanded by the ornately carved entry façade, featuring a fluted Doric portico with a broken pediment. Symmetrical flanking wings complete the composition. The garden façade is dominated by a long gallery that lightens the rear of the house with a series of arch topped bronze doors. Offering spectacular views of the great lawn, the Little River, and mature oak groves beyond, the transparency of the garden façade is in complete contrast to the solid mass of the front entrance façade.

First-floor rooms are designed to connect with adjacent exterior "rooms", such as the walled kitchen garden, the family terrace and several lawn parterres which transition into the landscape.

On the interior, elegantly proportioned formal rooms feature elements such as custom marble mantelpieces, robust moldings, an elegantly carved walnut–paneled library, and a stunning wood ceiling in the family room modeled after one in Luyten's Castle Drogo. On the exterior, bronze doors, mahogany windows, a graduated slate roof, Brazilian granite water-table, and lead-coated copper in combination with a refined limestone exterior, signal a truly timeless character.

As part of the overall landscape design, the folly serves not only as a visual focus but also as a destination point for visitors touring the grounds of this expansive estate. The dome and columns are of solid, load-bearing limestone resting atop a granite stylobate—all in keeping with the great classical architectural design of Wayside Manor.

A further image from this project can be found on page 34.

TURNING A CORNER:
THE IONIC CAPITAL
Unlike all the other Orders, the Ionic capital is not symmetrical on all four of its sides, which can create a problem. This drawback was resolved in antiquity by the use of a capital with a diagonal volute at the external corner, as shown above by architects Franck & Lohsen at Wayside Manor. This arrangement allows the volute to turn the corner, preventing the side face of an Ionic capital being placed next to front face of the adjacent capital. Similar examples found in antiquity are the Temple on the Ilissus and the Erectheum. Renaissance architect Vincenzo Scamozzi later proposed a version of the Ionic capital with four corner diagonal volutes, which in turn was later popularized by English architects Inigo Jones and James Gibbs.

"Timeless Beauty is always appreciated. From the elaborate to the simple, classical architecture enables people to experience man-made beauty."
Michael Franck

A Cotswolds Manor House

COTSWOLDS, ENGLAND

Robert Franklin Architect 科茨沃尔德庄园

THE EXISTING MANOR HOUSE in the center of this 200-acre wooded Costswolds property was built in the early 20th century to resemble a structure built some 300 years earlier, and was listed by the British Government as a building of architectural interest. However, although located on the top of a hill, the existing house was deeply flawed in the layout of its plan and failed to take advantage of the magnificent views in all directions. These shortcomings led architect Robert Franklin to successfully gain approval to radically alter and add on to the existing manor house—provided that it maintain a strong feeling of classical detail and style.

The gloomy core of the existing house was opened up, creating a new gallery and double-height entrance hall that features stone columns, robust classical detailing, an elliptical cantilevered stone stair, and wrought iron railings. Elsewhere windows were enlarged to fill the house with natural light, while also providing better views of the rural countryside.

The elevational geometry for the new west wing of the house is derived from the pure Pythagorean principles of the "golden rectangle," which dictate not only the overall composition but also the proportions of window panes. The addition, designed in the Palladian style, is separated from the body of the existing house via a "weak link," that prevents a collision of the differing classical styles. By designing the addition in a differing style, Robert Franklin cleverly allows both structures to be viewed independently, while also referencing many fine English country homes that have similarly been added on to, and evolved, over the centuries.

Finally, the impression from the entrance drive of approaching the back of the house, was reversed by giving more architectural detail to the gables, enlarging some windows and adding a domed and lanterned circular entrance lobby in a square shell with stone portico that breathed classicism, but is not a copy of any specific example. The result is a conflation of classical styles that is harmonious in its own right—and a stately manor house that finally lives up to the perfection of its setting.

SPECIALIST CRAFTSMEN: SYMM
For this project, architect Robert Franklin persuaded his clients to hire SYMM, one of the UK's leading building companies to oversee construction. Local Cotswold stone was used to construct the rubble walls and stone slates for the roof, and the inappropriate existing pointing was carefully raked out and replaced with lime mortar. SYMM is particularly known for its expertise in the manufacture and installation of cantilevered stone staircases—in this case, made from Lincolnshire Clipsham stone.

"Classic, of course, simply means best of its kind. So just how much breadth of taste can you accommodate within classical?"
Robert Franklin

A Georgian Country House

SOMERSET COUNTY, NEW JERSEY

Allan Greenberg Architect　乔治亚风格乡间别墅

THE CHARACTER OF THIS GEORGIAN COUNTRY HOUSE changes radically from the front entrance façade to the rear garden façade. Set on 100 acres of New Jersey's horse country, the clients wanted a house that not only looked as if it had been there for 150 years, but also stylistically related to the brick Georgian character of so many houses in the Short Hills-Bedminster area.

Dramatic and picturesque changes in massing and scale enliven the front façade, which are designed to hold the eye when approaching the house by car. The tall dominant cental section contrasts with the one-story-high colonnades that extend out at right angles and connect the symmetrical one-story-high office wing, on one side, and the garage/service wing on the other. The private garden façade is simpler in its detailing, and more Adamesque in its style. Here the scale of the exterior elements— like the octagonal bays and serliana windows—are a response to views back towards the house, from the gardens and forest, which is located a quarter of a mile away.

The plan was built up by creating several sequences of spaces that differ in size, shape, volume, and proportion. The principal axis begins at the front door and moves through a square vestibule, an elliptical hall with a low dome, a narrow gallery, a large rectangular living room, and into the garden. Different spatial sequences are revealed by entering through the two side doors, the octagonal pavilions, or the family entrance in the side arcades. The variety of these circulation routes is enhanced by views out through the large windows and French doors to the gardens, the distinct ornament and decoration of each room, and the arrangement of art and furniture.

All of this was possible due to the close collaboration between architect Allan Greenberg, decorator Elissa Cullman of Cullman & Kravis, and a team of experienced craftsmen.

A further image from this project can be found on page 32.

SIMPLIFIED ORDERS:
TOWER OF THE WINDS
With a single row of acanthus leaves surrounding a single row of palm leaves, this simplified version of the Corinthian order has become commonly known as the Tower of Winds Order—named after the ancient Greek temple that they adorned. Left half-buried, the temple did not receive attention until it was rediscovered and featured in Stuart and Revett's three-volume publication *Antiquities of Athens*. The use of the capital became popular in the 19th century used in the Federal and Greek Revival styles as a more restrained, yet still elegant, alternative to a Corinthian capital. Here architect Allan Greenberg has used the capitals, along with a pulvinated frieze and end scrolls, at the entrance portico of this Georgian Manor.

"Beautiful buildings lift the spirit, please the senses, and challenge the mind."
Allan Greenberg

River Oaks Townhouse

HOUSTON, TEXAS

Ike Kligerman Barkley Architects　橡树河别墅

THE CLIENTS OF THIS THREE-STORY TOWNHOUSE had previously lived in much larger 1930s Georgian Revival style house in the River Oaks neighborhood of Houston. By the time their six children had left home, the couple decided that they should find a smaller lot in the same neighborhood and build a not-so-large house. Their previous home contained many unique architectural elements that they had come to admire and, as such, instructed Ike Kligerman Barkley to incorporate these into the new design. There started a lengthy process of photographing, measuring, and cataloguing all of the items that they wanted to salvage and reuse. The clients also had a three-floor apartment in Chicago, and wished to emulate the spatially compact up-and-down layout. All the architects now had to do was combine the layout and salvaged architectural elements into a new design that would be appropriate for the historic setting of River Oaks.

Because the previous house had a Georgian rigor that disciplined the eclectic mix of salvage, it was decided that the new townhouse should also be Georgian in style. Ike Kligerman Barkley chose as a source of inspiration the 1808 three-story Nathaniel Russell House, an iconic Late-Georgian townhouse in Charleston, South Carolina. Overlooking the Buffalo Bayou, the new townhouse, like the Nathaniel Russell House, has a symmetrical street façade that is detailed with brick and a rusticated limestone base. Incorporated into the façade are elliptical windows recycled from the previous home.

Inside the townhouse, the principal rooms are located on the *piano nobile* and are reached by elevator or the grand elliptical stair which features an Art Deco style balustrade. Based upon a sketch from the client of a similar balustrade in Berlin, and fabricated locally in Houston, the staircase is illuminated by an antique stained glass window. Found by the clients in Philadelphia, it was designed by Jacques Gruber and exhibited at the Art Deco exposition of 1925 in Paris, and is a perfect example of the close interaction that the client had with Ike Kligerman Barkley in the design of this townhouse.

PRECEDENT:
NATHANIEL RUSSELL HOUSE
Built in 1808, the late-Georgian style townhouse of Charleston merchant Nathaniel Russell is a National Historic Landmark, and widely recognized as one of America's most important dwellings. Although architects Ike Kligerman Barkley referred to the famed townhouse for inspiration they made many changes, the most notable being the addition of the two close-set windows, united by an engaged Corinthian column, found on the front façade. The three-story elliptical staircase at the River Oaks Townhouse was also inspired by the staircase in the Nathaniel Russell House—but rather than being detailed in the Adamesque style, it is infused with Art Deco motifs discovered by the client while on their travels.

"Working in the styles of the past does not preclude personal expression. We look for the historic style that represents the best vehicle for the architectural story we wish to tell."
John Ike, Thomas Kligerman & Joel Barkley

A Regency Villa

CHESHIRE, ENGLAND

Francis Johnson and Partners 摄政风格别墅

THIS HOUSE REPLACES a Victorian farmhouse that was in poor condition and took no advantage of the beautiful views to the south over the Cheshire countryside towards the Welsh hills. It was built for a retired couple who had sold a large family house nearby and who now wanted a smaller house with gracious sized rooms.

The inspiration for the design was John Nash's campagna villas, particularly Cronkhill in Shropshire, which stands in a similar situation, closely allied to farm buildings. However, Cronkhill is nestled into the landscape and is only intended to be seen from one direction, whereas this house has been designed to be viewed from all four sides. By removing the hedges around the former farmhouse and defining the boundary to the south and east with a ha-ha, the new house now sits in a park-like setting amidst lush Cheshire pastures.

The interiors are detailed in the Regency style based on those at Cronkhill. The fluted door architraves have paterae at the corners cast from original 19th-century molds by George Jackson & Sons Ltd. of Sutton.

Although this house is the first revival of Nash's campagna villa to be constructed in the United Kingdom since the 19th century, the combination of formality and asymmetry, coupled with plentiful windows and generous loggias, make it a style well-suited to 21st-century requirements.

PICTURESQUE SPIRIT: JOHN NASH
Imitating the buildings in a Claude or Poussin landscape, Regency architect John Nash was a master in the Picturesque style, creating anglicized Italian vernacular villas in the English countryside. Amongst the best examples of this style are Sandridge Park and Cronhill—on which Francis Johnson & Partners based their design. Unlike more formal classical styles, this Italianate form of the picturesque is both restrained and romantic, and its success is dependent on creating a balanced asymmetrical design combined with a simplified material and color palette. Nash worked closely with the great landscape designer Humphry Repton on many of his commissions, and just as it was then, the collaboration between design professionals is paramount in the success of any project.

"Classical design can be a source of endless joy and fascination. If some of that pleasure comes through into the finished building, we can count it a success."
Digby Harris

A Georgian Manor

POTOMAC, MARYLAND

David Jones Architect 乔治亚风格住宅

AFTER TOURING A NUMBER OF EAST COAST HOMES, the clients of this Georgian mansion were drawn to Mount Pleasant—a country residence on the Schuylkill River outside of Philadelphia, built by a Scottish sea captain in 1761.

David Jones's strategy was to accommodate the family's modern program within a traditional Georgian five-part composition, with Mount Pleasant serving as the inspiration for the central two-story body of the house. The axial vista from front to back was maintained, while a new transverse stair hall provides connection to the wings and allows windows to flood the center of the house with light.

The challenge was to meet the extensive program while staying true to the modest proportions and character of the original. Flanking wings use arcades to mask the locations of the kitchen and party room, while symmetrical end pavilions, with garages below and guest bedrooms above, echo the central block—but manipulate floor levels to maintain a correct sense of proportion.

Using on-site measurements and detailed drawings from the United States Department of the Interior's Historic American Building Survey, David Jones meticulously reproduced and adapted much of the historic interior and exterior detailing found at Mount Pleasant. The front porch, Palladian window and third floor dormers were re-created. The robust palette of the exterior of the house—scored stucco on masonry, brick quoins and belt course, ashlar stone foundation, and true split shakes on open lath for the roof—also echo the original. Throughout the interior, detailing and trim were informed by architectural motifs found at Mount Pleasant.

Ultimately, this new Georgian Manor House delivers precisely what the owners were looking for: Mount Pleasant re-interpreted for a modern lifestyle.

A further image from this project can be found on page 9.

PALLADIO: AN AMERICAN PRECEDENT
Palladian architecture became popular in the Colonies during the mid-18th century, where a conscious effort for academic composition, balance, and symmetry became a prime concern. Adapted to a non-Latin climate there was an emphasis on central pediments; centralized Palladian windows; flat window arches with splayed bricks and keystones; a greater use of quoins and rustication; the use of a *piano nobile* or high basement; and elaborate doorways with arched fanlights, framed between pilasters or engaged columns, and surmounted by pediments. Many of the best examples of domestic Georgian architecture can be found in Philadelphia and its immediate neighborhood. Amongst these is the building that architect David Jones carefully studied and emulated, Mount Pleasant—which includes all of the listed Palladian attributes.

"Like architects before us, we are creating new and original designs while borrowing from the past with the purpose of moving architecture forward."
David Jones

A Georgian Manor House

VILLANOVA, PENNSYLVANIA

John Milner Architects 乔治亚风格庄园

HAVING LIVED FOR A TIME in Ireland, the clients developed a great appreciation for the distinctive Georgian-style manor houses that can be found in that country. The designs of Sir Edwin Lutyens and his interpretations of classicism were also on the minds of both the client and architect John Milner as he designed a house to capture those qualities while serving the needs of a modern family with eclectic tastes. Set on a rolling site amid formal and informal gardens, the house features vernacular materials merged with traditional Georgian elements and detailing. The carved limestone frontispiece and the turned limestone columns of the rear loggia are juxtaposed against rustic "dry-laid" fieldstone walls. Window and door openings and projecting hexagonal bays are trimmed with hand-molded red brick. Above the plaster cove cornice, the roof is covered with classic English plain clay shingles which have short exposures, cambered in their width, and finished with hogs back ridge tiles and bonnet hips.

The interior is organized by an enfilade that accommodates the principal staircase and provides access to the formal living room, dining room and library, as well as the informal family gathering spaces and kitchen. Adjacent to the formal gardens, the living room is dominated by a bespoke fireplace, also designed by John Milner, that features exuberant carvings in bold relief. The wood-trimmed walls of the dining room are inset with fabric panels and the plaster ceiling has a large recessed oval at its center. A columned veranda with a fireplace is a zone of transition between the family room and the informal gardens with a swimming pool and adjacent pavilions. The landscape is punctuated by architectural elements including a diminutive square stone tower and attached orchid house that frame the service court.

The defining characteristic of the house is the contrast between formal and informal, traditional and contemporary—manifested in the choice of building materials on both the interior and exterior of the house.

A further image from this project can be found on page 18.

GEORGIAN VERNACULAR: LUTYENS
Like many English Georgian country homes, this design by John Milner has little to no ornament on its exterior, other than the highly embellished limestone frontispiece—relying instead on a use of vernacular materials to provide definition, as well a proportioning system that, when used together, subtly reference the work of Sir Edwin Lutyens. The overall material choice of clay tiles, and local rubble stone outlined with red brick rustication, window surrounds, and string courses is taken from Lutyens' design for Millmead in Surrey. More noticeable, the front façade, dominated by the Baroque-inspired entrance surround and sweeping steps, is based upon that at Lutyens' masterpiece, The Salutation.

"Classical design was innovative at its inception, and still is."
John D. Milner

ART AND LIFE
NEW EDITION

Fifth Avenue Pied-à-Terre

NEW YORK CITY, NEW YORK

John B. Murray Architect 第五大道居所

ARCHITECT JOHN MURRAY was approached by a couple from Boston to create a pied-à-terre for their frequent visits to New York City. They had secured two rather mundane one-bedroom apartments in a pre-war building on Fifth Avenue, with the aim of transforming the combined space into something very special. The client's appreciation for ornament and detail encouraged a design where the entire apartment would be perceived as a hidden jewel within the city.

As with all of his designs, John Murray started the design process with a series of *analytiques* that closely looked at every detail and determined the amount of embellishment that would enrich each surface. Recurring themes and patterns were developed in wood, metal, stone, and plaster, as a way of shaping and unifying the entire space.

A rotunda-shaped vestibule connects to a vaulted gallery at the center of the apartment. The gallery opens to a beautifully proportioned salon and beyond to a library and master bedroom, each with distinctively shaped, coved ceilings. Pattern was used to enhance form. A stylized Greek key marches around a band in the entry dome, which draws the eye upward but then beyond the pattern to the apex of the dome giving a perception of additional height. As one looks along the hallway toward the master suite, the cadence of the stone pattern and banded, vaulted ceiling draws the viewer along the axis with a pause at a niche in the wall and then onward to another niche flanking a broad panel with art. These concepts are carried through the apartment with the idea of expanding the volumes to their best proportion. The overall impression of the apartment is that it has always been there—just as the client and John Murray intended.

Further images from this project can be found on pages 86 and 89.

ILLUSION: MANIPULATING SPACES
Paramount in apartment design is creating the illusion that the living space is larger than it actually is. Although part of this is dependent on the color palette, and the scale of the furnishings, simple architectural concepts also has a hand in manipulating how we perceive a space. In this pied-à-terre, architect John B. Murray has designed an open floor plan with enfilades that allow rooms to open up to one another; paneled walls and details are continued from one room to another, creating the illusion of one big space rather than several smaller ones; the embellished crown is designed as part of the paneled walls, allowing the now un-tethered ceiling to appear higher; baseboards are tall, and the chair rail is placed low on the wall tricking the eye into thinking the rooms are proportionally taller than they actually are.

"To this day the most modern of "classical" architecture exemplifies an appreciation of symmetry and attention to geometrical precision and proportion that visually equates with the highest of construction and design standards."

John B. Murray

Drumlin Hall

DUTCHESS COUNTY, NEW YORK

Peter Pennoyer Architects 冰丘庄园

DRUMLIN HALL is a modern-day version of a Federal style villa—a house that revels in perfect proportion, rigorous geometry, and the great flexibility of the classical idiom. Set in a valley of miniature hills, or drumlins, at first glance it seems to be an English house built years ago.

The exterior of the house, faced in warm buff sandstone, references a wealth of prominent Regency-style residences designed by the likes of Benjamin Latrobe, Sir John Soane, and S. P. Cockerell. Each of the four principal façades—each with its own distinct personality—delights in the possibilities of geometry, graceful proportion, and equilibrium.

The plan of the house revolves around two central axes and succinctly absorbs all of the requisite rooms into a contained rectangle with windows that express themselves symmetrically on the façades. Throughout, the attention to detail reinforces the classical essence of the house. In the stair hall, Greek Revival-inspired door casings, pilasters, and standing door pediments carved with anthemia mark the entryways set along the long axis while a set of faux-marble columns frame the library. Details, such as a repeated star motif, which purposefully hint to the client's Texan heritage, were added to personalize the design, and painted wood floors—in the stair halls, bar, and library—reference other important historic houses, such as Edgewater, in the area.

In designing the great curved stair, Peter Pennoyer looked to Cheekwood in Nashville, Tennessee—a house the client greatly admired. The scale and composition of the balustrade were refined to better fit the proportions of Drumlin Hall, and metalworker Jean Wiart was brought in to execute the delicate wrought pattern of garlands, pine cones, and eagles. In the upstairs stair hall, the most dramatic interior in the house, the firm incorporated niches for sculpture—as it did in the stair—fusing the art with the architecture.

Further images from this project can be found on pages 2–3 and 78.

INTENT: CREATING A SENSE OF PLACE
The design intent of Drumlin Hall was to successfully combine the owner's love of the rural Hudson Valley landscape, the architecture of Robert Adam, and the furniture of Duncan Phyfe. As such, equal importance was placed on the exterior architectural design, the positioning of the house in the pastoral landscape, and the design of the room interiors as a backdrop to display the owner's museum-quality collection of Federal-style furniture and Hudson River School paintings. As highlighted here, the balance between architecture, landscape, and interiors is the cornerstone of all classical design.

"The language of classical architecture gives us the framework to pursue our imaginations without making our work self-referential. Classicism is more than the canon of rules laid down by the masters such as Palladio."
Peter Pennoyer

A New Colonial House

PARADISE ISLAND, BAHAMAS

Hugh Petter, ADAM Architecture 新殖民风格别墅

LOCATED ON AN OCEANFRONT SITE on the northern beach of Paradise Island in The Bahamas, this house is designed in a Colonial style that borrows heavily from local traditional Caribbean details. The site is elevated 30 feet above sea level to minimize damage during hurricanes, whilst also maximizing the view out to the ocean from the pool terrace.

A wide two-story portico dominates the entrance façade, while a two-story veranda extends the full width of the house to take advantage of the ocean views. The pool terrace, with its open-air kitchen and dining area, offers a different type of experience, with a sense of enclosure and landscaped gardens.

The house built of concrete block and stucco with a shingle roof, provides over 10,000 square feet of modern family accommodation in a flexible plan. A double-height stair hall provides an impressive entrance to the central porticoed section of the house, while two subsidiary lower-story wings frame the formal front garden and contain a garage, office, games room, and gymnasium, with accommodation above. All wood trim, both internal and exterior, is made of mahogany, some of which has been painted to withstand the often harsh conditions. Rather unusually, many of the building elements were fabricated and assembled on the actual site, in order to avoid the Bahamas' high custom duties for importing construction materials.

Above the entrance door is a carved tympanum by the renowned architectural modeler and carver, Dick Reid—possibly one of the last examples of his work before he retired. The interiors are designed by Monique Gibson from New York, who worked closely with the owners and architect Hugh Petter to ensure a cohesive design.

A further image from this project can be found on page 94.

THE ORDERS:
ANTIQUITY & RENAISSANCE
Architect Hugh Petter has created a Colonial-style home that is dominated by a two-story pedimented portico on the entrance façade, and two-storey veranda on the oceanfront façade. Both use the same classical order throughout—a simplified Doric, almost indistinguishable from Tuscan, except for the pattern of ventilation holes that echo the sequence of mutules. The Doric columns on the first floor have a typical base, as illustrated by Vignola in the Renaissance; whereas the columns on upper story has no base at all, and are modeled after ancient examples such as those at the Theater of Marcellus in Rome. Tapered pilasters—which are quite rare in classical architecture—echo those used by English architects Inigo Jones and Edwin Lutyens on many of their commissions.

"Classical architecture enables each generation to invest their buildings with meaning and craftsmanship."
Hugh Petter

Ferne Park

DORSET, ENGLAND

Quinlan & Francis Terry Architects 弗恩庄园

FERNE PARK IS PERHAPS ONE OF THE FINEST modern-day examples of an English country house designed in the Palladian vocabulary. Skillfully reinterpreted for the location, the house was completed in 2002, and the following year won the Georgian Group's annual award for Best Modern Classical House.

Although Ferne Park is an entirely new construction, it is located on the site of a previous Georgian mansion that dated back to 1811, which had since been demolished, and as such, the local planning authority stipulated that any new house should be built of local stone, classical in design, and be no larger than the previous house. With the need for a classical design, the clients approached Quinlan Terry, whose work they were already familiar with.

Governed by a landscape that provides stunning views across the Wiltshire and Dorset countryside, Terry designed the house to have contrasting façades. The north façade, with dramatic views towards Wardour Castle, is designed with a pedimented portico of double-height engaged Composite columns and a monumental flight of stairs reminiscent of a grand Palladian Villa, whereas the south façade, which faces gentler pastures, is more intimate in scale featuring a summer terrace, rose garden, and a small entry flanked by Ionic pilasters. Inspired by Came House near Dorchester and Castletown Cox in County Kilkenny, Ireland, the design includes four local stones, as mandated—Chilmark stone for the ashlar face, Portland stone for the rusticated base and ornamental details, Upper Greensand (a local sandstone), and the durable York stone used for the steps and terraces.

Inside the house, the plan reflects the tradition of grand Georgian simplicity, with all the principal rooms opening off a central entrance hall. The staircase features cantilevered stone steps and a wrought-iron balustrade that leads to a Venetian window at the landing. Adding to the richness of the interior are fireplaces designed by Francis Terry.

Further images from this project can be found on pages 14–15 and 126–127.

DETAILS AND SCALE:
COMPOSITE ORDER
Typically, the Composite order is the richest of all the classical orders, combining the decorative acanthus leaves from the Corinthian order with the volutes from an Ionic capital. However, with each of the capitals on the north façade standing over 6 feet tall, and being located almost 50 feet above the surrounding terracing, architect Quinlan Terry elected to emulate the more restrained Composite capitals found in the pediment of Palladio's San Maggiore in Venice. By doing this, all of the carved detail is large enough to be seen and appreciated as guests ascend the grand entrance steps. For those experiencing the capitals from the adjacent upper floor windows, a large carved stone bee located between the volutes is an unexpected treat.

"Classical architecture is based upon using materials and methods of construction in such a way that they provide a durable and dependable building."
Quinlan Terry

A Greek Revival Country House

HUDSON VALLEY, NEW YORK

G. P. Schafer Architect 希腊复古风格乡间别墅

GIL SCHAFER'S DESIGN for this 6,000-square-foot new residence, in rural countryside just two hours' north of New York City was developed as a response to the clients' love of Thomas Jefferson's Palladian houses in Virginia. Sited in the middle of a large farm field overlooking the Hudson River Valley and the Catskill Mountains to the West, the house is designed to take advantage of the property's commanding views through broad openings placed along its western façade.

Wanting to ground the design in a regional aesthetic more appropriate to its size, Gil Schafer detailed the house as a hybrid of both Jeffersonian and Greek Revival precedent. The house has a five-part plan with a central block and wings separated by hyphens typical of late 18th- and early 19th-century Virginia houses, but it is detailed with the bold proportions and molding profiles of the Grecian-influenced houses built throughout the Hudson Valley in the 1830s and 1840s. Moldings throughout the interior are based on Greek Revival precedent.

The house's interior, designed in collaboration with New York interior decorator Miles Redd, is anchored at its core by a central entry hall with a sweeping curved stair and a large living room with double fireplaces made from black scagliola. A pine-paneled library and the master suite, featuring his and hers bath and dressing rooms, sit to one side of the central core of the house, and are balanced by an informal wing containing the family room, kitchen, and mudroom on the other side. Two guest suites are located on the house's second floor.

Developed by Warren Byrd of Charlottesville, Virginia, the landscape surrounding the house includes several terraces, hedged gardens, and an elegantly-proportioned swimming pool set between the family room wing of the main house and its carriage house.

A further image from this project can be found on page 33.

MAN-MADE MATERIAL: SCAGLIOLA
Developed as an affordable alternative to marble and other semi-precious stones, scagliola has its origins in antiquity, but it was not until the 18th and 19th centuries that it was employed on a grand scale in columns and fireplaces in some of the most important private homes of the time. Made from pigmented cast plaster, strands of raw silk can be pulled through the plaster mix to resemble veining, before being buffed with oiled felt or beeswax to create a marble-like surface. Architect Gil Schafer often uses this technique in his designs, as shown here by the pair of scagliola fireplaces at each end of the living room.

"The classical language of architecture has endured for thousands of years precisely because it can be reinterpreted and reinvented in every age, but it still retains fundamental principles of proportion and detail within each transformation."
Gil Schafer

Ashfold House

SUSSEX, ENGLAND

John Simpson & Partners 阿什福德庄园

ASHFOLD HOUSE was one of John Simpson's earliest commissions and has been hailed as a model for country houses in the 21st century. Set in the Sussex countryside, Ashfold was built as a small villa for a retired couple. It incorporates all the necessary modern comforts in a building that is not only beautiful but also economical. Its use of space belies its small size and its design draws inspiration from classical predecessors, particularly the work of John Soane.

Ashfold comprises two floors with a basement cellar. The first floor incorporates a large south-facing drawing room, and a freestanding conservatory. A top-lit tribune allows light to flood the second floor and down to the first floor entrance hall below. There are four bedrooms in each of the corners and the central section of the south side of the building houses a library.

Architecture has been used to define rooms and also to borrow space from other rooms. The breakfast room itself seems to have been carved from nowhere, occupying as it does a large bay window and the space beneath the stair landing. The dramatic lighting from the double-height bay window and the beautiful views over the parkland give an illusion of space beyond the size of house.

Much of this architectural illusion is made possible by the use of a modern sophisticated heating system, with a heat recovery boiler. The house's square plan ensures that it is compact to minimize heat loss. Mirrored shutters have been incorporated to further retain heat—just one of the many details that makes Ashfold not only a beautiful house, but also one that is economical to maintain.

Further images from this project can be found on pages 74 and 256.

VARIATIONS ON A THEME: SOANE
The inspiration for Ashfold House can easily be found in the work of Sir John Soane. Here architect John Simpson has carefully studied Soane's residential commissions, and created four distinct façades on a symmetrical Palladian square plan. The north façade of Ashfold recalls the triumphal arch entrance at Pitzhanger Manor; the east façade is dominated by a glazed bow window, an element often used by Soane; the south façade, faced in stucco with stenciled classical decoration, reflects that of Soane's own house in Lincoln's Inn Fields, London; and from the west façade projects a large conservatory, with fenestration reminiscent of the un-built conservatory at Pitzhanger. Uniting all four façades is the use of stylized pilasters, similar to those at Pell Wall, and many more of Soane's designs.

"So today, just as it was in the ancient past, durability lies at the heart of what architecture is all about."
John Simpson

A Neo-Classical Penthouse

SAN FRANCISCO, CALIFORNIA

Andrew Skurman Architects 新古典风格顶层公寓

THE OWNERS OF THIS DUPLEX PENTHOUSE apartment wanted a Classic yet modern design, which heavily influenced the architectural design by Andrew Skurman.

The architecture in the apartment is highly detailed, inspired by the homes of Robert Adam. Fluted pilasters, rosettes, and arches abound; this high level of classical detailing is both highlighted and contrasted by the stark white color in which it is painted; this apartment is what happens when complex 18th-century English forms meet modern sensibilities.

The entrance hall is regularized by the use of Doric pilasters inspired by the 4th-century Baths of Diocletian in Rome. A curved stair with a solid banister curves through the space, cutting into the strong rectilinear form to bring tension and movement to the space. Light bounces off of the many planes found in the detailing, flowing through a round porthole window made possible by the apartment's location at the penthouse level.

This location at the top of the building also allowed for a large laylight in the gallery and multiple fireplaces throughout the apartment. Arches in enfilade march along the two sides of the long gallery and break the otherwise monotonous hallway into a number of interesting spaces. The arches are utilized elsewhere in the apartment, marking the spatial distinction between the dining and living rooms. The kitchen is accessed through another arch, which can be closed off by a pair of mirrored French sliding doors. This creates an open view from the kitchen to the water beyond while allowing the host to close off the area when entertaining guests. By separating the spaces with large arches, the classical demarcation of distinct spaces for different functions was retained while allowing for a modern flow between rooms.

ENFILADE: **FRAMING A VIEW**
Derived from the French *enfiler*, "to string together as in a series," enfilades have long been a compelling way of opening up, and connecting rooms—physically and visually—without compromising the character of each individual room. A common feature in grand European homes, such as Versailles and Syon, enfilades can also be used to great effect in smaller projects, such as this penthouse apartment. Here architect Andrew Skurman has created a sequential revelation of architectural features connecting the master bedroom, entrance hall, gallery, and living room along a single axis that extends from one end of the home to the other. Although symmetry is implied, it is important to note that it is not the salient feature of enfilade.

"I believe in a different solution for each client within the unique harmony of Classicism."
Andrew Skurman

California Georgian

ATHERTON, CALIFORNIA

Eric J. Smith Architect 加利福尼亚乔治亚风格庄园

THIS CALIFORNIA RESIDENCE recreates the historic feeling of a symmetrical Georgian-style home, while maximizing the allowable building area, and also preserving several large and beautiful oak trees that anchor the property. By designing around the existing woodland gardens that enclose the site, the house appears to have been built decades earlier. Yet despite its classical detailing and sense of permanence, the house includes all the amenities of a modern-day home, and is highly engineered in order to comply with California's rigorous seismic requirements.

The weathered patina of the home is further enhanced by the selection of stone for the exterior. Sourcing aged-faced granite cut from the abandoned fissures of a Connecticut stone quarry, the clients selected only pieces that had been exposed to the elements for many years.

The "H"-shaped plan allows most rooms to have windows on two or three façades, maximizing the views out on to the planted gardens and lawns, while also allowing natural light to flood the house. The symmetrical plan provides clarity, with enfilades providing axial views from one room to another, and out towards the swimming pool and pool pavilion beyond. Collaborating with a team of highly-skilled craftsmen and artisans, Eric Smith custom-designed all of the millwork, cabinetry, and plaster moldings in the house. It is this careful attention to detail, combined with material selection, that creates the feeling of a much older home.

Further images from this project can be found on pages 12–13, 43 and 44.

ORGANIZATION:
SYMMETRY AND AXIS

The most distinctive characteristic in classical architecture is the use of symmetry and axis—both in organizing the plan and detailing the façade. Typically, as per this Georgian design by Eric J. Smith, the plan and exterior elevations are symmetrical either side of an axis, or centerline, starting at the front door and running through the building. Secondary axes within the house follow the circulation routes, as in this case where upon entering the entrance hall the primary axis, which leads to the living room, intersects with a secondary axis aligned with the stair hall. Either side of the living room on their own secondary axes are the library and dining room (each with bay windows centered in the room), which in turn creates a symmetrical garden façade.

"Using symmetry and balance, the embrace of natural light and the honest use of durable materials united by the craftsmen's hand—these are the ingredients for good Classical Architecture."
Eric J. Smith

Pietra Mar

NAPLES, FLORIDA

Smith Architectural Group 彼得拉玛别墅

THIS RESIDENCE IN NAPLES, Florida is reminiscent of homes of the Italian Renaissance in its scale, setting, and style. It is first seen while crossing a coral stone bridge. Its rhythm of segmented arches and piers reflect in the lake below. The home is raised on a plinth for practical as well as aesthetic purposes. Its bottom course of stone splays out as in ancient stone buildings. The first story is sheathed in coral stone with stucco above a stringcourse. A barrel tile roof adds to the Mediterranean style.

The plan is organized around a central courtyard, with a mosaic pool at its center. The two-story main section of the house is on axis with the pool and overlooks the Gulf of Mexico to the west. Arcades line each side of the pool and connect to the guest suites.

Bronze entry doors with antique gold hardware set the tone for the opulent interiors of the home. A curved staircase, accompanied by a bronze balustrade featuring gold detailing, gracefully ascends to the master suite. The stair hall is the pivot point and axis to the living room, where rich fabrics, paintings, and materials are all balanced and proportioned with the classical detailing of the ceiling coffers and Corinthian capitals. The capital's theme of dolphins is repeated on the exterior of the house.

The main hallway features a stone patterned floor that is reflected in the plaster groin vault ceiling. The colors are a muted palette that offset bronze lanterns and antique furniture and art. The Morning Room lightens the mood with a painted wood ceiling, country French furniture and *trompe l'oeil* caryatides framing arched French doors. The Pub Room's butternut paneling and embossed gilt and painted leather panels bring a masculine room to life, while a gothic vaulted ceiling gives an almost aristocratic touch.

A further image from this project can be found on page 23.

MATERIAL INFORMS STYLE:
COQUINA STONE
Sitting high above the Gulf of Mexico with a splayed plinth foundation, Pietra Mar is intended to emulate the Italian Palazzos found along the coastline of the town's namesake in Italy. Using natural coquina, an indigenous coral stone from North Florida which ages beautifully over time, the house is a departure from the stucco façades of the Mizneresque Mediterranean style, and instead evokes the Italianate heritage of Palm Beach homes such as Il Palmetto and Casa della Porta, designed by Maurice Fatio in the late 1920s. The use of coquina in the Italianate style produces a restrained yet imposing exterior in sharp contrast to the delicate richness of the interior decoration.

"We are traditionalists and we strive to make a house look like it has always been there."
Jeff Smith

Pentimento

NASHVILLE, TENNESSEE

Ken Tate Architect 旧影重现别墅

AT FIRST GLANCE, this residence in the historic Belle Meade neighborhood of Nashville, Tennessee, appears to be a modest Georgian-style home, in keeping with the scale of the other houses in the neighborhood. But this is just a trick of the eye, created by manipulating scale and mass. Here, Ken Tate opted to use fewer windows—three bays instead of the typical five bays that you would find on a house of this size—creating the illusion that that the house is not as wide as it actually is. Dividing the mass into smaller parts also helped to diminish the perceived size of the house. On closer inspection you experience the actual scale of the house where everything, although proportional with the overall design, is oversized in scale.

The design of this home is based upon the concept of vernacular progression—the manner in which a house grows and evolves over time. Disparate parts of the home resemble a vernacular Tennessee barn, a Georgian house, a Federal wing, and a Colonial Revival addition. In part due to Ken Tate's use of period appropriate materials—weathered antique slate, cedar shake, tin, local field stone, and painted clapboard—these attached structures appear to have been built over two centuries, and create an immediate sense of history and place.

The concept of vernacular progression is not mere window dressing, and was used throughout the house. In the areas that were supposedly added during the Colonial Revival era, the exterior walls of the original Georgian home suddenly become interior surfaces, creating rooms and hallways where stone, clapboard, sheetrock plaster walls, and glass all combine together.

The Georgian-style creates order and is the perfect setting for the home's formal rooms; the Federal portico appears to have been added in the early 19th century; and the rooms at the rear of the house reflect the Colonial Revival style, which being less rigid than the other classical styles is perfectly suited for more modern comfortable spaces such as the family room and kitchen.

MULTIPLE USES: **SHUTTERS**
Wood shutters are frequently thought of as a standard fixture on buildings—from ancient Greece to Renaissance Italy, and from Georgian England to the capes of New England and the plantations of the South. Historically they have been employed for a variety of reasons—controlling the amount of light and ventilation that enters a room; providing privacy; protection from the weather and unwanted intruders; and, since Victorian times, to merely enhance the aesthetics of a building. Here, Ken Tate references the work of Virginia architect William Lawrence Bottomley by having solid paneled shutters for the first-floor windows (held in place by graceful iron shutter-dogs), and louvered shutters above.

"We are now returning to the humanist order of the early Classical sources. This is being done by looking back through the lens of the Colonial Revival."
Ken Tate

A Georgian Country Estate

GREENWICH, CONNECTICUT

Wadia Associates 乔治亚风格乡村庄园

THE SCENIC BEAUTY of Greenwich's mid-country sets the stage for the formal grandeur of this magnificent country estate. The brick and limestone Georgian design was inspired by the owners' preference for something tasteful and conservative on the one hand, yet grand enough on the other such that it could readily accommodate the frequent large-scale entertaining required for the philanthropic and fund-raising work that the family is involved in.

In keeping with traditional Georgian style, the front façade of the house is elegant yet restrained, featuring relatively little ornamentation with the exception of limestone accents for the quoins, window surrounds and entrance portico. The rear of the house, however, is undeniably grand with long sweeping views of the lawn and formal garden from the back terrace. Centered between bay windows topped with circular balustrades, the rear portico is proportioned exquisitely, and is given further emphasis by limestone pilasters and columns that continue up to the second floor.

The interior of the house is equally as grand. The luxurious details of the principal rooms bestow upon them an enviable elegance. A cupola above the central staircase leading off the entryway bathes the sweeping stair hall in sunlight. To give the space added distinction, the balusters were custom designed and individually carved from mahogany. In keeping with the formal symmetry of Georgian architecture, the dining room and living room lead off either side of the stair hall. The dining room ceiling features delicate plasterwork in the Adam style. Many of the rooms in the house were fitted with antique stone fireplaces procured from England. Such was the case with the library, which also features custom-made burled oak paneling milled in England. The decorative motifs reflected in the paneling are recurring elements that can be found throughout the architectural design of the house. This continuity is essential to the symmetry of Georgian design, and provides the foundation for the refined elegance associated with this architectural style.

Further images from this project can be found on pages 28–29 and 57.

CHANGING SCALE:
PORCHES AND PORTICOS
The orders can be applied in a number of different ways to alter the appearance of a building. Here, Wadia Associates have introduced two-story giant Doric columns, which, given their size, allow the whole building to assume the character of that order. The use of brick, slate, and limestone further emphasize the heavy masculinity of this Georgian design. Dominating the façade, the use of the giant order in residential design can be credited back to Italian architect Andrea Palladio and the popularity of his *Four Books of Architecture*.

"Our pluralistic approach to classical design is not rooted in a slavish imitation of the past. More appropriately, it is based upon reinterpreting the past and updating it for the present."
Dinyar S. Wadia

"It is not enough to know your craft, you have to have feeling."
Edouard Manet

CREDITS

版权信息

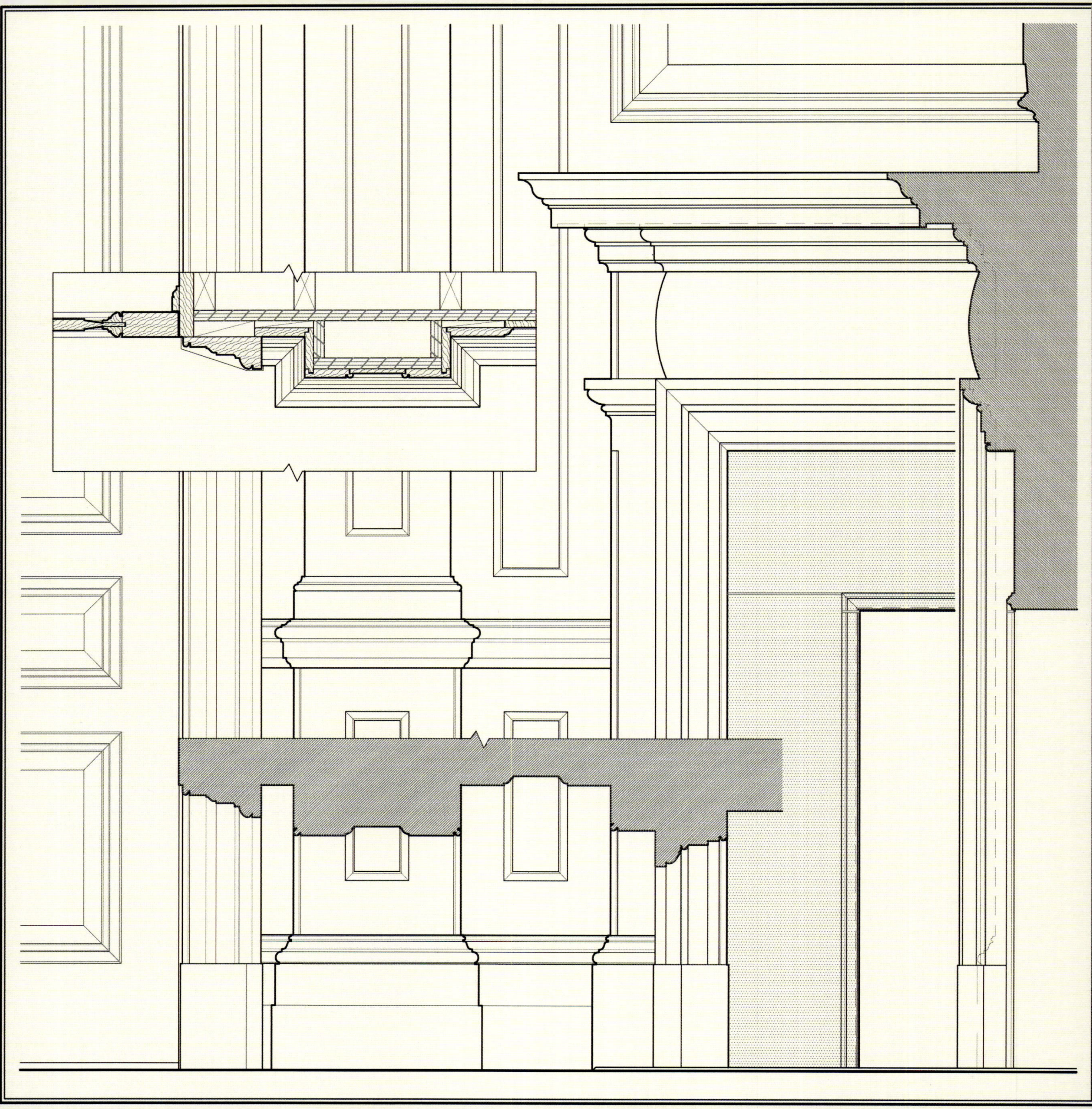

About the Author
关于作者

A NATIVE OF MANCHESTER, England, Phillip attended the prestigious Prince of Wales' Institute of Architecture in London—where he was taught by many of the architects featured in this book. He received a Degree in Architecture from his hometown university, before moving to America, where he gained a Masters in Architecture from The University of Notre Dame.

As a well-regarded expert on classical architecture and interiors, Phillip has worked for some of the most recognized residential architectural firms in America. He is a Fellow Emeritus at The Institute of Classical Architecture (ICA), and Classical America (CA), where from 2006 to 2009 he served as the jury coordinator for the annual Arthur Ross Awards. He has taught classical design at the ICA & CA and at Notre Dame, and has been a guest critic for the Interiors Departments at Parson's School of Design and the Fashion Institute of Technology.

Phillip is also the author of *New Classicists: Wadia Associates*, which features a foreword by HRH The Prince of Wales, and an Introduction by Paul Gunther—the President of the ICA & CA. He was also one of the contributing authors of the *Classicist: Volume VI* and served on the editorial team responsible for *A Decade of Art & Architecture 1992–2002*.

OPPOSITE
A design drawing for a mahogany panelled library, with a bespoke mantelpiece
Phillip James Dodd

Acknowledgements

致谢

ABOVE: Lawrence Voaides of Traditional Cut Stone

BELOW: Quinlan and Francis Terry

THE GOAL OF THIS BOOK has always been to feature the work of those scholars, architects, and craftsmen who have inspired me in my career, with the hope that the work features in these pages will have a similar effect on others.

Although all of those featured in the preceding pages represent the very best in their fields, and were selected purely on those merits, it is nevertheless very gratifying to be able to showcase the work of so many people that I have become friends with over the years. These friendships are not only based upon a mutual admiration of each other's talents, but also an unconditional willingness to offer help and advice when called upon. In particular I would like to thank Richard Carbino, David Tyrrell, and Lawrence Voaides for their support and friendship.

In addition to the scholars, craftsmen, and architects whose work is featured, I would also like to thank all of the photographers and illustrators who have so beautifully recorded all of the designs. In particular I would like to thank my friend Jonathan Wallen, whose photography I can finally include on a front cover.

I am especially grateful to David Easton for providing such a sincere and heartfelt foreword. I am humbled to have someone of Mr. Easton's stature in the design community show such interest in a young classicist, and his idea for a book. As a torchbearer of classicism for almost 40 years, Mr. Easton is one of the reasons that we are now experiencing a revival in Classical architecture, as evidenced in these pages.

In some respects this book represents a journey that has come full circle. As an undergraduate, thanks in part to HRH The Prince of Wales' A Vision of Britain, I was drawn to classical architecture and chose to write my dissertation on the work of Quinlan Terry. Not surprisingly this subject was frowned upon by the School of Architecture, and my paper was passed along to the History Department to grade, where the professor commented "this study contains all the things that I normally object to; it is personalized, it is biased, and it is opinionated. But what an excellently mature and convincing piece of writing in defense of an architect. I have become a confirmed Terryite, and find that much of this dissertation is of publishable quality." Well, it may have taken almost 20 years, but I can now finally claim to have a book published featuring the work, and writings, of Mr. Terry.

Moving on, I will always consider myself fortunate to have attended The Prince of Wales' Institute of Architecture, where I was exposed to, and taught by, so many great architects, artists, and craftsmen. I would particularly like to thank Robert Adam, Julian Bicknell, and John Simpson—all of whom have supplied both essays and projects that feature in these pages—for teaching me the fundamentals of classical design and detailing. I would also like to thank master carver Dick Reid, who taught me the importance of listening to, learning from, and working with craftsmen of all trades. I feel blessed to have been taught by, and inspired by, such a wealth of talent.

Special thanks are reserved for my two mentors Richard Sammons and Dinyar Wadia. Richard was the one that tempted me to make the journey across the pond to New York, and more importantly to stay and start a career in America. He took me under his wing, and personally honed my design skills. Dinyar then gave me the opportunity to put those skills into practice. He also provided me with an environment where not only my talents as a designer and writer could flourish, but also where I could become the person that I am today. I will always be indebted to you both.

I would also like to thank everyone at Images Publishing Group, including Alessina Brooks, Mandy Herbet and Driss Fatih, but especially Paul Latham, who could have approached anyone to write this book, but for some reason contacted me. I would also like to acknowledge my friends Tom Maciag and Robert Lominski for their support and assistance whenever called upon.

My final thanks are reserved for my family, as they have always been, and always will be, the source of my inspiration. I could not have asked for a more supportive and loving family—and it is this in life that I am most grateful of. To my parents, brother, and my dearest Theresa, I love you.

ABOVE: Dick Reid

BELOW: Dinyar Wadia

The spiral staircase in this library is fabricated in maple, and is modeled on a stair in the Chapel of San Lorenzo, Santa Fe.
Fairfax & Sammons Architects

Library
参考书

BY DEFINITION, CLASSICAL ARCHITECTURE borrows heavily from the past and, as illustrated in the preceding pages, it is critical that we thoroughly understand what has come before us. However, it is equally important that we see how these theories and practices are now being interpreted for contemporary classical designs.

Below is a list of 22 books that I keep close by for inspiration. By no means an exhaustive list, these volumes provide additional insight into classical theories, architectural details, historical precedent, and most importantly—the work of today's best classical architects.

Theory

Classical Architecture
Robert Adam
Harry N. Adrams, 1991

The Classical Orders of Architecture
Robert Chitham
The Architectural Press, 2006

Classical Architecture
James Stevens Curl
W. W. Norton & Company, 2003

Parallel of the Classical Orders of Architecture
Donald M. Ratner (editor)
Acanthus Press, 1998

Theory of Mouldings
C. Howard Walker
W. W. Norton & Company, 2007

Details

Manuale del Recupero:Del Comune di Roma
Maura Bertoldi (editor)
Tipografia del Genio Civile, 1997

The Architectural Treasures of Early America: Volumes I–XVI
Lisa C. Mullins (editor)
The National Historical Society, 1987

Plasterwork: 100 Period Details
Jeremy Musson
Aurum Press, 2000

Decorative Details
William & James Pain
A. Tiranti, 1948

Building Details
Franck M. Snyder
W. W. Norton & Company, 2008

Historic Precedent

The Domestic Architecture of Sir Edwin Lutyens
A. S. G. Butler
Antique Collectors Club, 2003

J. Neel Reid Architect
William R. Mitchell, Jr.
The Georgia Trust, 1997

The Work of William Lawrence Bottomley
William B. O'Neal & Christopher Weeks
University Press of Virginia, 1985

The Classical Country House
David Watkin
Aurum Press, 2010

The Houses of McKim, Mead & White
Samuel G. White
Rizzoli, 1998

Modern Interpretations

New Classicists: Wadia Associates
Phillip James Dodd
Images Publishing Group, 2007

Timeless Elegance
David Easton
Stewart, Tabori & Chang, 2010

New Traditional Architecture
Mark Fergsuon & Oscar Shamamian
Rizzoli, 2011

American Houses: The Architecture of Fairfax & Sammons
Mary Miers
Rizzoli, 2006

The Great American House
Gil Schafer III
Rizzoli, 2012

Peter Pennoyer Architects
Anne Walker
The Vendome Press, 2010

Radical Classicism: The Architecture of Quinlan Terry
David Watkin
Rizzoli, 2006

This hand-forged baluster was inspired by the staircase at the Hotel Dodun in Paris, constructed in 1726.
Gold Coast Metal Works

Contributors
供稿者

DAVID EASTON

David Easton is considered to be one of the world's most sought-after interior designers. He founded his own firm in 1972, and quickly gained recognition for his classically inspired, traditional interiors. In the 1980s, his opulent English-style interiors became emblems of the age. David was named to the Interior Design Hall of Fame in 1992, and has twice been presented with Classical America's Arthur Ross Award. Most recently, his peers honored hum with the Lifetime Achievement Award at London's Design and Decoration Awards. *Architectural Digest* has included David in their 'Top 100 Designers in the World', a record 10 times.

JEREMY MUSSON

A leading commentator on the English country house, Jeremy Musson is the author of several books including *English Country House Interiors*. He has written for *County Life* magazine since 1998 and was the architectural editor for nearly a decade. Jeremy also co-wrote and presented the acclaimed BBC television series on important country houses called *The Curious House Guest.*

PROFESSOR JAMES STEVENS CURL

James Stevens Curl is a Member of the Royal Irish Academy, a Fellow of the Societies of Antiquaries of London and of Scotland, a Fellow of the Royal Incorporation of Architects of Scotland, and the author of many highly acclaimed books, including *A Dictionary of Architecture and Landscape Design*, *Victorian Architecture: Diversity & Invention*, *Georgian Architecture: The British Isles 1714–1830,* and *Freemasonry & the Enlightenment: Architecture, Symbols & Influences.*

DAVID WATKIN

David Watkin is Emeritus Professor of the History of Architecture at the University of Cambridge, an Honorary Fellow of the Royal Institute of British Architects, A Fellow of the Society of Antiquaries, and Vice-Chairman of the Georgian Group. His many books include *Sir John Soane: Enlightenment Thought and the Royal Academy Lectures*, *Morality and Architecture Revisited*, *Radical Classicism: The Architecture of Quinlan Terry*, and the seminal *A History of the Western World.*

ROBERT CHITHAM

Robert Chitham is an architect who has specialized in building conversation as well as neo-classical design. He was the first Directing Architect of English Heritage after its formation in 1983, and was subsequently President of ICOMOS UK. He is the author of *Measured Drawing for Architects*, and the *Classical Orders of Architecture*.

HENRY HOPE REED

For well over half a century Henry Hope Reed has been the foremost spokesman for the classical tradition in architecture and its allied arts. In 1968 he co-founded Classical America, the pioneering organization that promoted the current resurgence in classical and traditional design. His works include the popular *The Golden City*, *The New York Public Library: Its Architecture and Decoration*, and *The U.S. Capitol: Its Lesson for Today.*

Outstanding craftsmanship means using traditional techniques, whether that be hand-carving or forging steel on a blacksmith's anvil. These are labor intensive and highly-skilled processes for which there is no substitute.
Chesney's

QUINLAN TERRY · QUINLAN & FRANCIS TERRY ARCHITECTS · ESSEX, ENGLAND

As one of the most celebrated architects practicing today, Quinlan Terry has played a leading role in the revival of classical architecture, emphasizing traditional materials, construction methods, and ornament as valuable solutions to modern architecture. Together with his son Francis, who regularly exhibits drawings at the Royal Academy, their firm continues the architectural firm that was started by Raymond Erith in 1928, specializing in Palladian inspired designs. Quinlan has been honored with numerous awards, including the Richard H. Driehaus Prize for Classical Architecture.

ROBERT ADAM · ADAM ARCHITECTURE · WINCHESTER, ENGLAND

Robert Adam's contribution to the classical tradition is internationally acknowledged, as a scholar and as a designer of traditional and progressive classical architecture. He trained at the University of Westminster and in 1973 won a Rome Scholarship. Robert co-founded the practice *ADAM Architecture*, the largest traditional architecture practice in the world. He works on a diverse range of projects including major private houses, historic building extensions, public and commercial buildings, masterplanning, speculative housing and has pioneered objective coding. His work is widely published and he lectures both in the UK and overseas.

JULIAN BICKNELL · JULIAN BICKNELL & ASSOCIATES · LONDON, ENGLAND

Regarded as one of today's leading classicists, Julian Bicknell founded his own firm in 1984 after running the project office at the Royal College of Art for six years. Julian's architecture is both pragmatic and theatrical, as he seeks to combine the advantages of modern technology and mass-production with occasions for pleasure, surprise, and fine craftsmanship. His formal designs are based on rigorous modular and geometric systems applied both in the setting out of plan, section, and elevation and also throughout the detailing. These systems are designed in the first instance to order and simplify the process of construction—but provide as a by-product the calm and enviability of classical architecture and interiors.

JOHN SIMPSON · JOHN SIMPSON & PARTNERS · LONDON, ENGLAND

Established in 1980 and based in the Bloomsbury area of London, John Simpson & Partners specializes in designing in the classical and vernacular tradition. Their portfolio of work encompasses urban masterplans, public and educational buildings, private residences, and interior design. John was instrumental in the New Classicism movement of the 1980s, and is rightfully acclaimed for his counter proposal for the design of Paternoster Square in London. Amongst his built projects is an addition to Gonville and Caius College at the University of Cambridge, and the Queen's Gallery at Buckingham Palace. He is the recipient of many awards, including the Arthur Ross Award for Architecture in 2008.

PETER PENNOYER · PETER PENNOYER ARCHITECTS · NEW YORK CITY, NEW YORK

Since its founding in 1991, Peter Pennoyer Architects has built a substantial and varied body of work for residential, commercial, and institutional commissions often involving significant historic buildings across the country. The study of history is the generating force of the firm, and each project, from restorations and renovations of historic properties to new construction, unites vigorous scholarship with an inventive reinterpretation of the classical language. Peter currently serves as the Chairman at The Institute of Classical Architecture & Art, and is the co-author of numerous architectural monographs including *The Architecture of Delano & Aldrich*, and *The Architecture of Warren & Wetmore*.

JOHN B. MURRAY · JOHN B. MURRAY ARCHITECT · NEW YORK CITY, NEW YORK

For over 15 years John Murray's office has been lauded for its traditional aesthetic, its simplicity of form, and in particular, its supreme craftsmanship. Starting with hand-drawn plans, the firm's designs all reflect their context, while seamlessly integrating sophisticated and unique details. This old world care, and respect for the creative clarity that comes from a visual as well as in-person dialogue between client and architect, are the firm's hallmark. Over the years John has collaborated with many of America's top decorators, and is considered an expert on the Park Avenue vernacular.

RICHARD CAMERON · ATELIER & CO. · BROOKLYN, NEW YORK

Richard Cameron founded Atelier & Co. in 2009. The firm takes its name from the old French word for an artist's or craftsman's studio, and practices architectural design as a fine art in the traditional way, collaborating on projects with other architects, artists, and craftsmen—hence the "& Co." Previously Richard was the Design Director of Ariel—The Art of Building, a firm that he co-founded with David J. Cohen. Perhaps his greatest achievement is as co-founder of the Institute of Classical Architecture, where he has taught every year since its inception in 1992. At present he serves as Vice-Chairman of the Institute, as well as Director of its Beaux-Arts Design Atelier.

HUGH PETTER · ADAM ARCHITECTURE · WINCHESTER, ENGLAND

Hugh Petter, a director of ADAM Architecture, is a well-recognized authority on classical and traditional architecture. He is currently working for The Duchy of Cornwall on the masterplan and code for a sustainable urban extension to the town of Newquay. Hugh won Rome Scholarships in 1990 and 1991 before returning to London in 1992 to establish the Foundation Course at The Prince of Wales' Institute of Architecture, where he served as Senior Tutor for six years. He has published essays on various architectural subjects, and is a visiting tutor to several colleges in the UK and overseas.

AIDAN MORTIMER · SYMM · OXFORD, ENGLAND

Aidan Mortimer is Group Chief Executive of SYMM, a Fellow of the Society of Arts, a Brother of the Art Workers Guild, and has a master's in Historic Conservation. His building company SYMM, was established in 1815 and is based in Oxford, England, with offices in London and New York. The company specializes in the restoration and alteration of historic buildings and also creates bespoke joinery and stonework in its Oxford workshops. SYMM works worldwide and past projects include work at Buckingham Palace, The Houses of Parliament, and every one of the University of Oxford's colleges.

ALAIN OLIVIER · GOLD COAST METAL WORKS · HUNTINGTON BAY, NEW YORK

President and founder of Gold Coast Metal Works, Alain Olivier is a self-taught expert in the art of decorative architectural ironwork. By establishing his company's atelier with the workmanship of experienced artisans in Olomouc in the Czech Republic, he has helped re-enliven the storied craft. GCM designs, fabricates, and restores architectural metalwork for fine residences throughout the world. Alain has an MBA in Finance from the University of Michigan and a bachelor's degree in Modern European History from the University of Chicago. Before founding his company in 2002 the New York native worked in commodities.

PAUL CHESNEY · CHESNEY'S · LONDON, ENGLAND

Paul Chesney, founder of Chesney's, was educated at Bradfield College, Berkshire and went on to read law at Fitzwilliam College, Cambridge. He practiced as a solicitor before deciding in 1984 to start a company that now employs over 250 people and produces over 5,000 marble and limestone chimneypieces every year. The business has developed massively since it started, initially concerned solely with restoration of reconditioned antique fireplaces, but now known for being the first company to make high-quality reproductions in marble and stone, widely available at an affordable price. Chesney's has offices, showrooms and warehouses in the UK and the USA.

KEVIN CROSS · KEVIN CROSS STUDIO · REDDING, CONNECTICUT

Kevin Cross is the owner and founder of Kevin Cross Studio, an art and design studio that offers traditional painting and gilding services, and specializes in creating custom furniture, decorative art, and lacquer work for interior designers, architects, and private clients. The company also provides restoration services for collectors of antique furniture and decorative objects. Established in New York City in 1987, Kevin Cross Studio is now located in Redding, CT.

RICHARD CARBINO · TRADITIONAL CUT STONE · TORONTO, CANADA

Richard Carbino founded Traditional Cut Stone with his partners David Tyrrell and Lawrence Voaides. Specializing in handcrafted masterpieces in limestone, marble, and sandstone, their company is home to the largest number of classically trained stone carvers in North America. The 20,000 square foot workshop provides both stonecutting and stonecarving services, and is home to their art department, which provides sketches, renderings, and plaster models of many of their designs.

FOSTER REEVE · **FOSTER REEVE & ASSOCIATES** · NEW YORK & LOS ANGELES

Foster Reeve, president of Foster Reeve & Associates, holds an MFA from Parsons School of Design. His passion for design led him in 1992 to found his own company to singularly pursue the craft of traditional plasterwork. Foster's mission is to ignite interest in plaster in the design community and promote its use as the best material choice for quality trim and decoration. He specializes in custom designed plaster moldings, ornamentation, and bas-relief, as well as integral color stuc pierre, scagliola, and a host of decorative wall finishes. His work adorns major projects around the world.

JIM MATTHEWS · **H. G. MATTHEWS BRICKWORKS** · BUCKINGHAMSHIRE, ENGLAND

H. G. Matthews Brickworks was first established in 1923 by Jim's grandfather. Still a family-owned and operated company, it is a small old-fashioned works that has retained many traditional skills and techniques. This has enabled them to thrive as suppliers of hand-made bricks to the conservation sector, as well as for high-end new homes. Recently, with assistance from Colonial Williamsburg in Virginia, the company has relearned the technique of wood firing, which means authentic bricks, with beautiful wood glazing are once again available for the repair of buildings predating the 18th century.

TIMOTHY BRYANT ARCHITECT · NEW YORK CITY, NEW YORK

Since 1997, Timothy Bryant Architect has created residences that celebrate craftsmanship, beauty, and sense of place. A Classicist by training, Timothy was born and raised in England, where he was introduced to architecture and the building arts at an early age by his father, who was a builder. Whether ground-up construction, restoration and renovation, historic preservation, interior decoration, or adaptive reuse, each project embraces the challenges of contemporary living, integrating the demands of sustainability and energy efficiency with time-tested construction methods and fine materials.

CURTIS & WINDHAM ARCHITECTS · HOUSTON, TEXAS

Since the beginning of their collaboration in 1992, Bill Curtis and Russell Windham have found a mutual interest in producing traditional architecture that combines a deep respect for architectural history and context with a well-studied knowledge of traditional and classical detailing. Although best known for their work in Houston's historic suburbs, the firm has completed projects throughout America. Bill and Russell have both been recognized nationally in numerous publications, and are the recipients of several architectural awards.

FAIRFAX & SAMMONS ARCHITECTS · NEW YORK CITY, NEW YORK

Partners in business as well as in marriage, Anne Fairfax and Richard Sammons established their award-winning architectural firm over twenty years ago. Their body of work reflects theories of proportion and order that have been passed down through scholarship and practice for generations. Both principals are sought-after critics, teachers, and jurors in the academic community, with Anne serving on the board of The Institute of Classical Architecture & Art, as well as The Royal Oak Foundation, and Richard internationally recognized as an expert in the field of architectural proportion.

FERGUSON & SHAMAMIAN ARCHITECTS · NEW YORK CITY, NEW YORK

Mark Ferguson and Oscar Shamamian built their firm on a shared enthusiasm for traditional architecture that was nurtured while they both worked at the celebrated decorating firm of Parish Hadley. Since its founding in 1988, the firm's core practice has been the design of custom residences, including city apartments, suburban houses, and country estates. Working with some of today's most recognizable decorators, such as Michael S. Smith and Bunny Williams, the firm has been widely published and has been named by *Architectural Digest* as one of their 'Top 100 Architects and Designers.'

The Tempietto at Settrington continues the tradition of the Georgian gentleman creating his own private arcadia of a classical temple in the park.
Posthumously executed from the design of Francis Johnson

FRANCK & LOHSEN ARCHITECTS · WASHINGTON, D.C.

As winners of the prestigious Arthur Ross Award for Architecture in 2011, Franck & Lohsen Architects specialize in classical architecture and town planning. Founded by Michael M. Franck and Arthur C. Lohsen, the firm has earned distinction for its timeless and elegant designs, often with buildings or regions that are highly sensitive to their historic nature. With projects all over the United States as well as projects in Italy and England, their work displays a thoughtful and sophisticated combination of classical approaches and modern sensibilities.

ROBERT FRANKLIN ARCHITECT · OXFORDSHIRE, ENGLAND

Robert Franklin established his practice in 1978, and kept it deliberately small so that he might remain involved in each aspect of every design. Advocating that architecture should be a pleasure, and not a stylistic crusade, Robert's designs are the result of long and affectionate study of the best examples from the past. All design work is done by hand that allows a continuous development of details unique to each project. He is listed on the RIBA's Accreditation Register at the highest level as a 'Specialist Conservation Architect,' and is recipient of numerous design and conservation awards.

ALLAN GREENBERG ARCHITECT · WASHINGTON, D.C.

Allan Greenberg established his firm in 1972, and now has offices in Greenwich, New York, and Washington, D.C. The firm has an international reputation for combining contemporary construction techniques with the best architectural traditions to create solutions that are both timeless and technologically progressive. Allan's articles, teaching, and lectures have exerted a strong influence on the study and practice of classical architecture, and in 2006, he was the first American to be awarded the Richard H. Driehaus Prize for Classical Architecture.

IKE KLIGERMAN BARKLEY ARCHITECTS · NEW YORK, NEW YORK

With an emphasis on residential design the work of John Ike, Thomas Kligerman, and Joel Barkley, is articulated in a variety of styles synthesizing historical precedent with contemporary vision. Since its inception in 1989, the firm has earned international recognition for their rich vocabulary of form, and their considered approach to detail, material and craft. The firm has earned numerous design awards, and has been featured among *Architectural Digest's* 'Top 100 Architects and Designers' since 1995.

FRANCIS JOHNSON AND PARTNERS · YORKSHIRE, ENGLAND

Chartered Architects Francis Johnson and Partners was founded in 1937 by the late Francis F. Johnson, who was regarded as one of England's pre-eminent classical architects. The practice, now overseen by senior partners Digby Harris and Malcolm Stather, is best known for its Country House design—although they also design smaller houses and cottages, as well as commercial buildings, and interiors. The practice has received numerous awards over the years including, most recently, the 2009 Giles Worsley Award for a New Building in a Georgian Context from the Georgian Group.

DAVID JONES ARCHITECTS · WASHINGTON, D.C.

For over 30 years, working from a studio atop a townhouse in Washington's Dupont Circle neighborhood, David Jones Architects designs houses that fit throughout the Washington area and beyond. David's background includes undergraduate and professional degrees from Princeton, and study at King's College, Cambridge. His work is characterized by a sense of authenticity, attention to detail, and a particular mix of the traditional and the contemporary. The firm maintains a national reputation, regularly receiving feature articles in numerous publications.

JOHN MILNER ARCHITECTS · CHADDS FORD, PENNSYLVANIA

John Milner Architects specializes in the evaluation, restoration, and adaptation of historic buildings, as well as the design of new homes that reflect the rich architectural traditions of the past. The firm's firsthand experience with the history and construction technology of three centuries of American architecture informs its approach to both historic preservation and new design. In addition to his professional practice, John has taught at the University of Pennsylvania's Graduate School of design for over thirty years, receiving the Perkins Award for Distinguished Teaching in 2007.

Select details, in a variety of materials, from an American country house
Wadia Associates

G. P. SCHAFER ARCHITECT · NEW YORK CITY, NEW YORK

G. P. Schafer Architect is a small New York City-based firm specializing in classical and traditional residential architecture. In each of its projects the firm emphasizes a commitment to quality, craftsmanship, rigorous detail based on an understanding of historic precedent, and most importantly, comfort and livability. Led by principal Gil Schafer III, the firm has been widely published in books and periodicals on both sides of the Atlantic and has been the recipient of numerous awards including three Palladio Awards for outstanding traditional residential design.

ANDREW SKURMAN ARCHITECTS · SAN FRANCISCO, CALIFORNIA

Andrew Skurman founded his firm in San Francisco in 1992. As principal and owner of Andrew Skurman Architects, he focuses on superbly-crafted custom houses that are perfectly and logically planned to the specific requirements and wishes of his clients. His expertise lies in the elegant and refined expression of Classical architecture and the interpretation of French, Georgian, and Mediterranean styles. Andrew has received the honor of being named a Chevalier des Arts et des Lettres by the Minister of Culture of France.

ERIC J. SMITH ARCHITECT · NEW YORK CITY, NEW YORK

The office of Eric J. Smith Architect reflects Eric's personal values, those of tradition, timelessness, grace, and a commitment to quality. For over 20 years, Eric and his staff have been designing high-end, custom residences for clients in the United States and overseas, working in a variety of settings, styles, and architectural vocabularies. The firm is well known for its collaboration with some the America's finest interior designers, landscape designers, and craftspeople—and in particular the work it has done over the years with David Easton.

SMITH ARCHITECTURAL GROUP · PALM BEACH, FLORIDA

Since opening his office in 1989, Jeffrey Smith has established a practice that is committed to the pursuit of classical architecture. Following in the tradition of Palm Beach's celebrated architects of the past, Jeff's work displays a refined elegance coupled with exquisite detailing and superb craftsmanship. He subscribes to the idiom that the rebirth of tradition yields architecture with meaning. Much respected in Palm Beach, he has served as Chairman of the Town's Architectural Commission and Landmark Preservation Commission.

KEN TATE ARCHITECT · NEW ORLEANS, LOUISIANA

Ken Tate started his own firm in 1984 and has since become one of the most sought-after designers in his native South. Ken's understanding of classical architecture is enhanced by his love of vernacular styles, and while some of his designs offer a faithful representation of historic styles, others marry elements from several periods to give the impression that they grew and changed over time. By employing this design technique, as well as the use of traditional craftsmanship, all of his homes have a romantic feel that makes them appear to have been lived in for generations.

WADIA ASSOCIATES · NEW CANAAN, CONNECTICUT

For over 30 years, Dinyar Wadia has earned a reputation for designing classically-inspired homes, gardens, and interiors. His finely detailed residences display a remarkable versatility and adaptability within the classical language, and are characterized by a passion for excellent detailing, use of fine materials and exceptional workmanship. Dinyar's core design philosophy is to emphasize the integral relationship between a home and its gardens—an approach that has seen him garner numerous architectural and landscape design awards.

Photography & Drawing Credits

照片及绘图版权声明

PHOTOGRAPHY:

PAUL BARKER: *172-175, 250*

GORDAN BEALL: *176, 226–229*

MICHAEL BIONDO: *27*

IAN BRADSHAW: 34, 156, 158

NICK CARTER: *5, 62*

ROB CHATTERSON: *11, 121*

PAUL COSTELLO: *33, 202–207*

JOHN CRITCHLEY: *64, 66, 67, 90, 93, 128–131, Rear cover*

SCOTT FRANCES: *194–197*

JEFF GAHRES: *114, 140–143*

JASON GRIMES: *88*

MICK HALES: *48*

HESTER + HARDAWAY: *21, 144–147*

PETER KINGSFORD: *58*

ERIC KVALSVIK: *9, 177–179*

ANDREW LAWSON: *14–15, 126–127, 198, 201*

DAVID DUNCAN LIVINGSTON: *12–13, 44, 212–215, 216–219*

DIANNA NELSON: *110, 113*

PETER OLSON: *239, 240 bottom right*

LISA ROMEREIN: *40, 152–155*

KIM SARGENT: *23, 24, 54, 105, 222–227*

DURSTAN SAYLOR: *10, 20, 22, 32, 51, 85, 148–151, 164–167, 168–171, 184–187, 238–239, 242*

TIM SOAR: *240 bottom left*

ROBERT STAVELEY: *30*

SIMON UPTON: *38, 136–139*

JONATHAN WALLEN: *Front Cover, 2–3, 16, 17, 19 below, 28–29, 57, 77, 78, 102, 104, 116, 188–193, 230–235, 244, 252*

MATT WARGO: *18, 53, 180–183*

NEIL WAVING: *199, 200*

DRAWINGS & RENDERINGS:

RICHARD CAMERON: *86, 89*

PHILLIP JAMES DODD: *238*

CHRIS DRAPER: *94*

CARL LAUBIN: *Inside Dust Jacket, 37, 68*

JAMES STEINMEYER: *8*

MANUEL TAN: *43*

FRANCIS TERRY: *56, 61, 63*

COURTESY OF:

JULIAN BICKNELL & ASSOCIATES: *26, 52, 71, 72, 132–135*

CHESNEY'S LTD.: *25, 106, 109, 246*

FRANCK & LOHSEN ARCHITECTS: *47, 157, 159*

ROBERT FRANKLIN ARCHITECT: *161, 162*

H. G. MATTHEWS BRICKWORKS: *125*

JOHN B. MURRAY ARCHITECT: *82*

PETER PENNOYER ARCHITECTS: *81*

HUGH PETTER: *241 top right*

FOSTER REEVE & ASSOCIATES: *118, 120*

JOHN SIMPSON & PARTNERS: *74, 208–211, 256*

SYMM: *19 above, 96, 99–101, 122, 160, 163, 255*

TRADITIONAL CUT STONE LTD.: *117, 240 top left*